Real-Time Indoor Location Determination Solutions for tracking shoppers in Retail

Kevin Curran

Table of Contents

1. Introduction 4

2. Location Based Services for Retail 8

3. RFID Tracking Solutions...11
   3.1 CoreRFID - RTLS...12
      3.1.1 Impinj Speedway UHF RFID xPortal 13
      3.1.2 ThingMagic Astra integrated UHF RFID reader & antenna 14
   3.2 3M 16
   3.3 Redbite...18
   3.4 Odin Technologies Ireland 19
   3.5 Trolley Scan 20

4. WiFi Tracking Solutions 23
   4.1 Ekahau..25
   4.2 Trapeze Networks 29
   4.3 AeroScout...32
   4.4 Motorola Proximity Awareness and Analytics 34
   4.5 Insiteo 35

5. Bluetooth Tracking Solutions 36
   5.1 BlipTrack 36
   5.2 Direct enquiries - Indoor Tracking System 38

6. Mobile Tracking Solutions 39
   6.1 Path Intelligence 39
   6.2 Nearby Systems 41
   6.3 LocationLabs 43
   6.4 GloPos 44

7. Indoor Map Based Systems 45
   7.1 Point Inside 46
   7.2 Google Indoor Mapping 47
   7.3 Wifarer 49
   7.4 Qubulus..51
   7.5 SenionLab..53
   7.7 Walkbase...55
   7.8 Indoor Atlas,...57

8. Miscellaneous Tracking Solutions 59
   8.1 Ultra-wideband 59
   8.2 Ultrasound Positioning..62
   8.3 Wireless Sensor Networks – Zigbee 63
   8.4 Infra-RED (IR) Tracking Solutions 64

8.5 Camera Based 65

   8.5.1 Shopper Tracker...66

   8.5.2 ShopperTrak...67

   8.5.3 Prism Skylabs 68

9. Visitor Footfall Capture 69

  9.1 Experian FootFall...69

  9.2 IN4MA 70

  9.3 CheckCount 70

  9.4 Axiomatic Technology Ltd 71

  9.5 Euclid...72

10. Mobile Loyalty..74

  10.1 Shopkick 76

  10.2 Near Field Communications (NFC) 77

   10.2.1 NFC and Windows 8...78

   10.2.2 Google Wallet 80

   10.2.3 NFC Security Issues 81

11. Conclusion...83

Appendix A: RFID Presence Marketing Typical Deployment...........................86

Appendix B: WiFi Presence Marketing Typical Deployment 88

1. Introduction

Shopping centres are starting to seek to maximize at least Key Performance Indicators (KPI's) such as the total number of shoppers coming to the centre (i.e. 'footfall'), the number of times that shoppers come to the centre over the course of a year (i.e. 'repeat visits') and the length of time that shoppers stay at the centre (i.e. 'dwell time'). Many shopping centres have technical systems in place to capture information on the total number of shoppers in the centre. These are usually camera-counting solutions that count the total number of shoppers entering and exiting the building. Centres also may have the ability to estimate repeat visitors.   Some centres have installed license plate recognition software in their car parks and hence have the capability to interrogate this dataset and identify the number of times the same car has returned to the centre.   However, until recently it has not been possible to quantify dwell time - except through sporadic, one-off surveys – and so 'dwell time' has only played a limited role in evaluating a centre's performance.

This book is focussed on looking into the various commercially available location tracking technologies that may prove a best fit for tracking visitors in large indoor public spaces which allow clients to ascertain footfall, repeat visits and dwell time. The systems which are potentially suitable must be capable of "Presence Identification". That is the identification of a customer upon entry in a physical retail environment and "session-metrics" which is the shopping pattern of a customer in a physical store including time of entry/exit. Two main technologies are most suitable for this - active RFID and WiFi. Active RFID systems installed at the entrance of shops are able to conclusively identify a card/tag carrying customer upon entry. The customer would not have to produce a card if the card was present on a shopping trolley. The read range of the portal would be adequate if placed near a doorway. In order to capture session metrics an active RFID system would also capture the exit of the customer through the portal. WiFi localisation systems would also track an active tag or mobile device but would not require the installation of portals at doorways.

There are overviews of relevant systems which use varying wireless technology standards such as RFID, Bluetooth, WiFi along with related new tools which combine social media. Quotes for recommended tracking systems and contact details are included. While cost is an important aspect of recommending a people tracking system integrator, in this case proven experience has been judged to have more weight. Hence the order of precedence associated with the recommended technologies. Facebook Places, Foursquare and just about every Social Networking seem to be focusing on location these days. The problem however is they all rely on some sort of check-in, whether it is manual check-in by the user or using the latest GPS/Geo-fencing based check-in. The Indoor map based systems in section 7 are all quite new. They offer some very interesting technologies for the near future but those systems are mainly being rolled out overseas and there is very little use of them in the UK. Some of them rely on crowdsourcing which is where the public and other commercial entities upload their indoor maps to the companies' servers for other people to use.

It seems that the most likely systems to implement are Active RFID or WiFi 802.11 Real time location services (RTLS). We recommend Active RDID however and the rationale is outlined later. There are various Active RFID providers. A number were contacted but one stood out and they are CoreRFID. In reality, CoreRFID were quick in responding to emails and offer a quite competitive product. They also have an evaluation kit which may work well for initial trials. 3M among others for instance proved to be more difficult to reach each time and took much longer to reply. They also seem to have more expertise in rolling out healthcare solutions. Likewise Odin Technologies have not had much experience in the retail sector. They are more comfortable dealing with manufacturing and larger scale systems. Having said that, I am fully aware of how professional their services are and the attention to detail which they possess in the actual system roll-out. All the companies contacted were reluctant to give definite prices even when I would specify the typical square footage of a likely trial site and the amount of shops involved. This was not surprising however as most sites will involve dealing with the IT infrastructure already in place and radio frequency interference can differ widely from site to site. One recommended technology option is to go the route of Active RFID portals. This allows the portals to be moved from location to location with relative ease. There is still an effort involved in integrating the 'readings' from each portal but the installation of WiFi portals should allow the devices to communicate with a portable server on site maintained by the Card Group. This approach is to be recommended over the WiFi 802.11 RTLS approaches. All those approaches require tight integration with the retailer's core Wireless infrastructure. It also demands a through mapping of each retail unit which can take quite an amount of time. Both approaches will rely on shoppers carrying tags but the Active RFID approach wins out on being the simpler approach.

Case studies are highlighted throughout the report for all relevant technologies. The visitor tracking solutions outlined in section 9 are quite close and in fact ideal. Sensewhere are one such company. They require little outlay on capital but the unfortunate aspect is that they simple count 'people' and there is no way to 'tag' these people to allow the in-depth demographic tracking of shoppers as hoped for. The only way to identify shoppers is through a. tags b. vision systems, c. mobile applications and d. mobile devices.

Tags can be passive or active and range in price accordingly. They are required for the Active RFID solutions (and sometimes the WiFi RTLS solutions) but a person carrying one who has previously opted in can then be tracked and identified. Camera based systems for the most part are very expensive, quite intrusive and overkill for a shopper tracker system. The ethics involved could turn into negative publicity for a company. Mobile applications as outlined in section 7 and section 11 allow the customer to be tracked but they must install the application, keep it turned on and the actual mapping is reliant on services which do not seem to have come to the UK yet. It is one to watch in the next 12 months. Finally, mobile devices such as phones can be tracked by WiFi systems when the WiFi on the smartphone is turned on or they can of course be tracked by systems such as Path Intelligence. In the case of path intelligence, the visitor is anonymous but in the case of using WiFi the person could be identified if they have opted in. Here however the system needs to be integrated with the retail units networking infrastructure. As mentioned later, Path Intelligence are launched their *Flockr* product which allows the visitor to be identified but it seems to be more a mobile coupon service. They have said however that it does lead to customer identification.

A recommended approach is to deal with a third party who scopes the initial trail, provides the equipment, tags etc. and works with the company installing the system to ensure customers can be tracked at an acceptable level. Further rollouts can then perhaps be taken care of by the customer and pitfalls encountered in the initial rollout should be preventable in the other site rollouts. The likely costs for any site are unlikely to be less than £5000 UK. Again, the installation of portals which are portable will then reduce this for other trials should the equipment not be installed permanently. The most obvious additional benefit of the recommended Active RFID tracking system that will let a company offer additional product or service ideas to clients is to offer superior visitor analytics. As mentioned in section 9.5 where a case study for the Euclid system is mentioned, a retail owner with multiple shops could begin to see more clearly visitor dwell times and act on outliers noticed in their different outlets. There is also the chance to offer loyalty promotions to clients through a system which can identify repeat customers. This may be an added value service useful to many shops. Appendix A provides a short graphical overview of what an Active RFID rollout might look like and Appendix B shows a typical Wireless 802.11 (WiFi) rollout and the likely problems of having some dead zones.

2. Location Based Services for Retail

Location-based services have already grown into a billion dollar industry, and since 70% of mobile calls and 80% of data connections originate from indoors, having a more precise positioning technology that works indoors will become increasingly more important. For retail areas in particular, the advantages of understanding the path that shoppers take is significant. Such information can assist the owners to:

- Evaluate and improve their retail tenancy mix by identifying which stores shoppers group together on a typical shopping trip;
- Understand the foot traffic to each retail tenant;
- Ensure stores are being charged the optimal rent given their location within the centre;
- Identify underutilized areas within the shopping centre;
- Understand the impact of anchors stores on the centre;
- Measure the implications of particular promotions or centre events;
- Assist with planning day-to-day centre management operations, such as cleaning and security.

Nokia are beginning to trial an indoor location based system (not yet available commercially) which is based on Bluetooth. The Nokia is now in trials in stores where Beacons throughout a store that leverage Bluetooth 4 can send signals to mobile devices with Bluetooth 4 and identify a user's location to within 10 centimeters. Devices that have Bluetooth 4 include the iPhone. With that level of precision, a vendor could present the mobile user with a promotion on a specific product when it is right in front of the user, and the offer could be targeted to shoppers based on past purchases or other factors. When the customer reached the checkout stand, the discount could be applied automatically. A system with this precision could allow a new range of indoor navigation systems to be built.

Mobile phone based tracking from Path Intelligence is implemented in a number of sites including Promenade Temecula in southern California and Short Pump Town Center in Richmond, Va. Both malls alert customers via small signs that they can opt out of tracking by turning off their cell phones. Forest City says that they are not looking at any one particular shopper, but they are trying to determine information on a larger scale. That information may include shopping behaviours to see how many people stop at a particular product display at a Nordstrom and how many people actually purchase that product. Of course, the downside to such a powerful system is that all the mobile phones are anonymous.

Others which I examined and spoke to include Nextnav[1] have offices in the US and India. Rather than concentrating on the device itself, they are instead concentrating on the actual physical network around urban areas. They install carrier grade transmitters in a given area, that emit their own proprietary signals that can be used my mobile devices to calculate their position both horizontally and vertically. It could fill a void as a solution for large scale projects, but it seems too large scale for simply tracking people in small urban indoor areas. They also only claim a horizontal accuracy of 20m, which is not ideal for places like shopping centres.

GeLo[2] are based in Holland, and have a similar approach to NextNav in that they are aiming to produce beacons that will enable positioning on the device without the need for GPS. Instead of using cell towers though, their hardware sends out signals via the new Bluetooth 4 standard. The backbone of the GeLo system will be an extensive database that drives the information and use of the GeLo beacon system. While a beacon transits its serial number, the mobile device will use this to access the appropriate information required by the particular application and user. Local data will be stored on the mobile device for network free access. When traveling to a new GeLo site, data will need to be preloaded or accessed through a network. It all looks very early days judging by their website with not much for developers to experiment with.

Micello[3] create and provide indoor maps off all kinds of venues round the world including airports, shopping malls, museums, stadiums etc. The maps are interactive, and can incorporate points of interest and routing as well. Apparently if there is anywhere you want mapped, you can just send them a request and they will do it for you. It would be interesting to see how quickly they can produce something and their pricing model.

Finally Visioglobe[4] are another french company who specialise in 3D mapping, both indoors and outdoors. They take mapping information from Navteq or

---

[1] http://www.nextnav.com
[2] http://www.gelosite.com/index.html
[3] http://www.micello.com
[4] http://www.visioglobe.com

via CAD files and create textured three dimensional maps that can incorporate routing and POIs. They provided the map for the Dubail Mall app which has won awards. Using swipe gestures you can pan, zoom and rotate, all in 3D. There is routing and POI search on there. They do work with other providers of indoor positioning systems but their product is overkill for simply tracking customers inside an enclosed area.

3. RFID Tracking Solutions

Radio Frequency Identification (RFID) is a technology which is in widespread use in areas like asset management and stock control. Radio signals are transmitted between a reader and a tag. An RFID tag consists of an antenna, a transceiver and a small amount of memory. An RFID reader has more functionality than a tag and in addition to an antenna and a transceiver it also contains a power supply, a processor and an interface to connect to a network. The tags may be either active or passive. The passive tags have no power supply and are activated by the signals scanning them. The active tags have a small power supply and this enables them to have a range of several meters when compared to less than 1 meter for most passive tags. RFID tags enable positioning by placing the readers at doorways or other such points of human movement. The network can then track people when the tag they are carrying, passes through a doorway. This information can be sent by the reader to a central server which can display the tag's location graphically. Active RFID tags have a much higher signal strength, as opposed to passive tags with a low signal strength that are depending on the RFID readers to power them. Active tags for a project tracking people within shopping centers could be set to have a range of approximately 20-30 meters which should give shop level visitor information. An active RFID tag has a unique identifier which can be continuously tracked with approximately 1 update per second as long as its signal reaches an RFID reader. The range of an active RFID tag is up to 200 meters. In order to save battery life and reduce the number of updates sent to the system, tags that only broadcast when moved can be used. Active tags however cost significantly more than passive tags. Therefore RFID tags must be handed out and linked to individuals or otherwise be available to the shoppers and collected again when leaving the centre. This requires shoppers as well as the centre to take action for the system to work, thereby lowering the penetration of the system.

Communications from active tags to readers is typically much more reliable than from passive tags due to the ability of active tags to conduct a "session" with a reader. Active tags, due to their on board power supply, also may transmit at higher power levels than passive tags, allowing them to be more robust in "RF challenged" environments with humidity and spray or with dampening targets (including humans, which contain mostly water), reflective targets from metal (shipping containers, vehicles), or at longer distances: generating strong responses from weak reception is a sound approach to success. In turn, active tags are generally bigger, caused by battery volume, and more expensive to manufacture. Many active tags today have operational ranges of hundreds of meters, and a battery life of up to 10 years. Active tags may include larger memories than passive tags, and may include the ability to store additional information received from the reader. Semi-passive tags, also called semi-active tags, are similar to active tags in that they have their own power source, but the battery only powers the microchip and does not power the broadcasting of a signal. The response is usually powered by means of backscattering the RF energy from the reader, where energy is reflected back to the reader as with passive tags. An additional application for the battery is to power data storage. Semi-passive tags have greater sensitivity than passive tags, possess a longer battery powered life cycle than active tags and can perform active functions (such as temperature logging) under its own power, even when no reader is present for powering the circuitry. Whereas in passive tags the power level to power up the circuitry must be 100 times stronger than with active or semi-active tags, also the time consumption for collecting the energy is omitted and the response comes with shorter latency time. *The battery-assisted* reception circuitry of semi-passive tags leads to greater sensitivity than passive tags, typically 100 times more. They have the ability to extend the read range of standard passive technologies, to read around challenging materials such as metal, to withstand outdoor environments, to store an on-tag database, to be able to capture sensor data, and to act as a communications mechanism for external devices.

3.1 CoreRFID - RTLS

CoreRFID has over 14 years' experience in the technologies that support track, trace, audit and control applications. Technology provided by CoreRFID is used to support applications in areas as diverse as asset management, logistics & delivery Tracking and security & access control. CoreRFID's customer base includes organisations like the BBC, Capita, Nokia, Thames Water and Norwich Union. CoreRFID has strategic partnerships with providers of Ultra High Frequency components, making it possible for CoreRFID's clients to exploit this technology.

The systems outlined here will require UHF tags. UHF tags offer good read ranges and the ability for a number of tags to be read simultaneously. It is best to buy tags from the same suppliers as the readers as they will have experience with previous clients as to which work best. Some leading UHF tag suppliers are Confidex, Xerafy and Omni-ID. ETSI tags are for the UK. CoreRFID provide a

Xerafy UHF Metal Tags Test Pack (see Figure 2) which consists of 14 UHF Class Gen 2 tags. The pack costs £47 and is a good way of determining which tags work best. This is especially useful if placing on metal shopping trolleys. Each Metal Tag Test Pack includes 2 pieces each of the following Xerafy tags: PICO X II, NANO X II, MICRO X II, Cargo Trak, Versa Trak, Global Trak and Data Trak.

3.1.1   Impinj Speedway UHF RFID xPortal

The Speedway xPortal (see Figure 1) is an integrated reader / antenna / housing for portal applications It builds on the capability of the Speedway Revolution family by packaging a reader together with the necessary antennas in an easy to install casing that allows for portal installations in office buildings, retail premises or stock rooms. The xPortal exploits the Speedway Revolution's adaptive configuration facilities to deliver superior read performance. Power-over-Ethernet allows single cable connectivity, further simplifying installation.

It broadcasts at frequencies 865-868MHz (ETSI), uses wired ethernet (10/100), supports EPC Global Class 1 / Gen 2 ISO18000-6C tags and has "Autopilot" adaptive configuring including Max throughput modes. The S/W platforms for interfacing with it is Windows (.Net), In fact, the Speedway xPortal uses the LLRP standard - XML messages passed to and from the reader on a TCP port so developers can develop on any platform that supports TCP based sockets. The Speedway xPortal is usually configured with Power-over-Ethernet.

Price: UHF reader £1,725   24v AC power (optional) £55

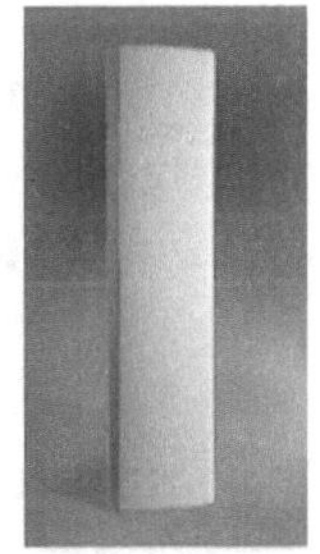

Figure 1: Impinj Speedway UHF RFID xPortal

Figure 2: Xerafy UHF Metal Tags Test Pack

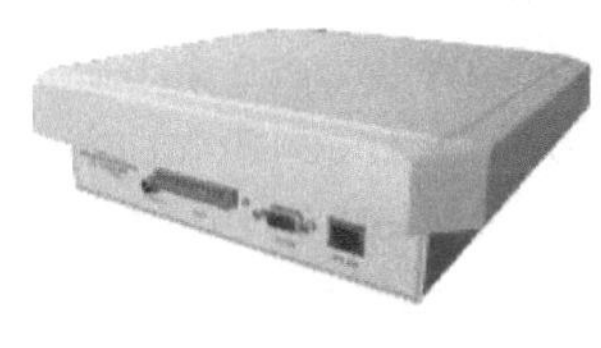

Figure 3: ThingMagic Astra UHF RFID reader & antenna

3.1.2 ThingMagic Astra integrated UHF RFID reader & antenna

ThingMagic's Astra integrated UHF RFID reader & antenna (see Figure 3) offers an enterprise level, high performance, low cost EPCglobal Gen2 RFID solution. Astra is ideal for commercial and enterprise environments that require unobtrusive deployments with a minimum of cabling, readers and antennas. The Astra includes network management and security features, such as DHCP for configuration and firmware management, and SSL/SSH-based security. Astra supports power over Ethernet (POE), or when AC-powered is available with an option for Wi-Fi backhaul.

It broadcasts at frequencies 865-868MHz (ETSI) and has 1 external antenna port plus integral antenna. The host interface is wired Ethernet (10/100) or WiFi (802.11 b/g). The S/W development platform is Java, .Net (C#). The max tag read rate is over 190 tags/second and max tag read distance is just over 30 feet (9M) with integrated 6 dBi antenna (36 dBm EIRP) parts.

Price: UHF reader £1,125   AC power and WiFi module £75

Applications to control the Astra reader, and all ThingMagic Reader products, can be written using the high level MercuryAPI. The MercuryAPI supports Java, .NET and C (for on-reader applications) programming environments. A fuller description can be found in this Developer's Guide[5]. The MercuryAPI Software Development Kit (SDK) contains sample applications and source code to help developers.

3.1.3 CoreRFID Development Kit

CoreRFID have a *Development kit / entry level system for active tag based Real Time Location System.* This active tag based real time location system identifies in real-time where a tag is. CoreRFID claim that the readers determine the position of tags through the timing of return signals the RTLS allows the location of tags to an accuracy of about 1m.

This package can be used for development of an entry level or pilot system. It provides 6 RTLS reader / anchor units (one with Ethernet), 6 power adaptors, mounting kits and tripod stands, 10 RTLS asset tags. Accesssories include

---

[5] http://www.corerfid.com/rfid%20shop/thingmagic/MercuryAPI_ProgrammerGuide_Jan12.pdf

power adaptors and mounting kits for reader / anchors and access to downloadable SDK, demo programme, user's manual, programmer's manual and installation guide.

The reader / anchor family allow the creation of a series of location cells; the Ethernet connected unit acts as the systems' master. The read range is up to 100 metres between tag and reader/anchor. In addition, CoreRFID software solutions are developed using the Microsoft .Net Framework making it easier to integrate track, trace audit and control applications with other back office systems.

Pricing
Price for Reader and 100 Active Tags: £5000

Contact: Jim Ryder   E   jim@corefrid.com or sales@corerfid.com
Address: CoreRFID Ltd. Dallam Court, Dallam Lane, Warrington, WA2 7LT, UK
Phone: +44 (0)845 071 0985   **Direct Dial:** +44 (0)1925 425 917
Site: http://www.CoreRFID.com

3.2 3M

3M have a proven track record in delivering tracking systems throughout the world. 3M's Track & Trace business is built around systems. These systems, whilst consisting of a number of component parts, are - in the vast majority of instances - sold as solutions rather than disparate hardware and peripherals. 3M can however provide an RFID based solution for the tracking of people. Their RFID based solutions utilise High Frequency (13.56 MHz) RFID. Seeking a solution where people 'walk' through a portal requires an Ultra High Frequency RFID and/or possibly Active/Powered RFID. Here in a nutshell is what they can provide:

A 3M RFID tag (which is guaranteed for the life of the file) (see Figure 4) is attached to a shopping trolley or any other device. They are not that expensive. The tag is then tracked around the centre by passing by strategically placed portals (see Figure 5). These portals are able to track multiple files all at once. They connect to existing PCs and feed information on the person's location to the 3M file tracking software (see Figure 6). Every PC in the systems is able to use the software to see the location of a person.

Figure 4: Tag

Figure 5: Reader

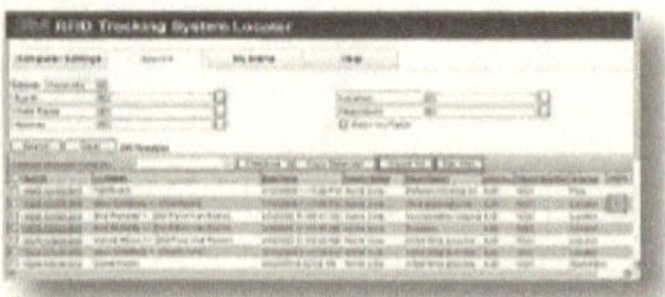

Figure 6: Tracking Software

3M tend to operate through value added resellers in this area, whereby 3M provide all hardware and software to a third party who deliver a solution including consultancy and user training. The prices shown below are indicative for hardware and software only. User training on the system is essential in order to fully recognise benefits.

3M Software and 3M provided interface: £28,315.00   (This is an enterprise license for unlimited users on the same network
3M RFID Tracking Portals: £1500.00
RFID Tags (1000):   £210

Contact: Chris Millican, Business Development Manager, 3M Track and Trace Solutions
Address: 3M Centre, Cain Road, Bracknell, Berkshire, RG12 8HT.
Site: http://www.3m.co.uk/patientrecordstracking
Mobile: 07767648089   **Work:** 0800 389 6686   **e-mail**: cmillican@mmm.com

3.3 Redbite

Redbite recommend utilising UHF C1G2 passive tags and fixed readers. The readers and antennas would be mounted in the ceiling and positioned appropriately to identify the movement of people in and out of shops (again subject to site survey). The solution can be delivered with a traditional architecture consisting of local web based application servers, an enterprise license and separate client licenses together with appropriate support or on a SAAS basis. However given the number of fixed readers involved they have assumed the former.

Budgetary estimates are:

1. 2 Fixed readers and antennas at £2,500 each = £5,000
2. Application software configured and bespoke to fit use case at £15,000
4. Professional services (to provide consulting, customisation and training) estimated at £7,500
5. Integration with back end database estimated at £10,000
6. RFID Printer / encoder at £3,500
7. Labels / Tags –at 40p each

Total (excluding tags) = £40,900
The professional services and integration may well come in lower but would need to be specified in more detail to be more accurate.

Company: RedBite Solutions Ltd, St Johns Innovation Centre, Cowley Road, Cambridge CB4 0WS, UK
Phone: +44 (0) 1223 421613

Contact: Chris Evans, Sales Director: +44 (0) 7930 327 778   or Alex Wong alex.cywong@redbite.com
Site: http://www.redbite.com

3.4 Odin Technologies Ireland

ODIN is a world-wide RFID company that has more than 500 successful RFID projects under its belt and over $50 million of R&D invested in a patented RFID-specific software. ODIN have a strong track record of high accuracy, scalability and integration with third party back-end systems. ODIN is one of the leading RFID companies in UHF. ODIN is headquartered near Washington, DC with offices in Boston, MA, Budapest, HU, Dublin, IRL and full-time resources in Toulouse, France and Geneva, Switzerland. The Dublin office is who I mainly dealt with.

Fixed Infrastructure
Portals: £4,800 per portal at a doorway or along a hallway etc.
EasyTAP 331: £2,500
EasyTAP reader licenses: £500 per reader (there are two readers in each portal)
Intelligent Asset Management (I AM) Software: £9,500 per centre/site

EasyTAP is a locationing engine that manages the reader infrastructure and turns the data into meaningful asset movement information for an application layer to consume. AM is the software application layer, and is an enterprise web solution. There is a base license of £47,500 which must be combined with site licenses, but Odin hold this pending the outcome of a given pilot.

Services - £20,000 - £30,000 for services, along with shipping, travel etc.

Total (approx.) for pilot: £22,000

A more detailed costing would of course have to be carried out were Box-it to proceed. Odin have carried out this type of use case – document tracking – for US Government agencies among others.

Contact: Ronan Wisdom
Phone: +353 1 443 4190
Email: RWisdom@odinrfid.com

3.5 Trolley Scan

The RFID-radar[6] system from Trolley Scan is an RFID based location determination system that they claim can track up to fifty tags and locate their location within a few seconds. The system has three main components, the reader, the antenna array and the tags (See Figure 7). The reader measures the distance of the signals from the tags from the antenna array which also energises the tags. The RFID radar works by measuring the distance the signal travels to two of the antenna then calculating the angle from each and movement can be detected by repeating this process. Trolley Scan determines the range of the transponder based on its received transmission. The reader has a location accuracy lower than 0.5 m, a pointing accuracy of 1 degree and can cover a maximum range of 100 m depending on the tag used with the reader. RFID-radar takes a relatively long time to determine the exact position therefore it is better suited to static situations where transponders are relatively stationary.

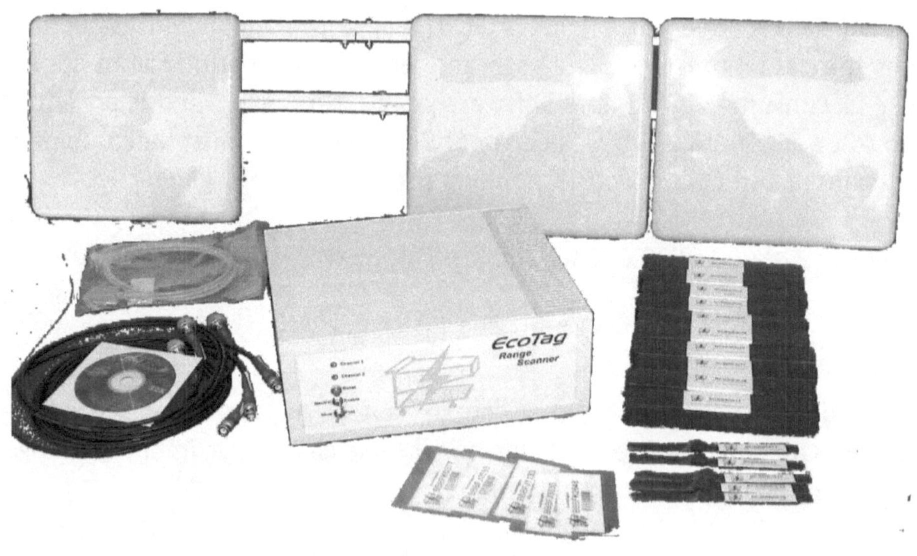

Figure 7: RFID Radar out of the box contents

A problem with RFID systems is that they need an external antenna which is 80 times bigger than the chip. Further, the present costs of manufacturing the

---

[6] http://www.rfid-radar.com/

inlays for tags have inhibited broader adoption, but as silicon prices are reduced and more economic methods for manufacturing inlays and tags are perfected in the industry, broader adoption and item level tagging may make RFID both innocuous and commonplace much like Barcodes are presently.

The RFID-radar by Trolley Scan, makes two measurements on each signal received from each transponder in its receiving zone: a range measurement and an angle of arrival. The angle of arrival measurement is virtually instantaneous and used in conjunction with range gives a 2D positioning system from a single measuring location. It also measures range with narrow bandwidth (10 KHz). RFID tags come in three general varieties: passive, active, or semi-passive (also known as battery-assisted).

Passive RFID tags have no internal power supply. The minute electrical current induced in the antenna by the incoming radio frequency signal provides just enough power for the CMOS integrated circuit in the tag to power up and transmit a response. Most passive tags signal by backscattering the carrier wave from the reader. This means that the antenna has to be designed both to collect power from the incoming signal and also to transmit the outbound backscatter signal.

The response of a passive RFID tag is not necessarily just an ID number and the tag chip can contain non-volatile, possibly writable EEPROM for storing data. Passive tags have practical read distances ranging from about 10 cm up to a few meters, depending on the chosen radio frequency and antenna design/size. The lack of an onboard power supply means that the device can be quite small: commercially available products exist that can be embedded in a sticker, or under the skin in the case of low frequency RFID tags. The transponders are low cost Tag-Talks-First type devices, capable of being produced cheaply. Trolley Scan also works with the Transponder-Talks-First Protocol which is suitable for fast moving tags which send their ID as soon as they have enough energy. The interference area is smaller than using RTF (Reader-talks-first) Protocol because the tags are less powerful than the receiver.

Due to very low operating power, Ecochip tags can be mounted on either side of objects as they will still be able to operate even when the large losses caused by the differences in dielectric constants of the objects in the path are taken into account (see figure 8). The range is reduced dramatically from the 13m air situation to around 4 to 6m depending on materials.

Figure 8: RFID Radar Supplied Ecotag claymore semi-passive RFID tag

The CR2032 battery has a life of approximately five years. The power level is 0.6 µW and the minimum range is 5 m as transponders might overload and stop if they are too close to the energizing field. The maximum range is 40 m. The Claymore tags (figure 8) have been developed to address the issue of sensitivity when attached to hard objects. The tag is mounted in a block of plastic which has a metal backing so that objects behind the tag do not influence the sensitivity. This means that the radiation pattern of the tag is more directional as it is intended to be attached to a hard object and does not need to radiate behind this tag. The long range stick Ecotag claims operating distances of at least 30m however when it is placed close to a hard surface, its performance degrades as the hard surface influences the sensitivity (the stick Ecotag has the radiation pattern of a conventional dipole). RFID readers that are in charge of the tags of an area may operate in *autonomous mode* (as opposed to *interactive mode*). When in this mode, a reader periodically locates all tags in its operating range, and maintains a presence list with a persist time and some control information. When an entry expires, it is removed from the list. Frequently, a distributed application requires both types of tags: passive tags are incapable of continuous monitoring and perform tasks on demand when accessed by readers. They are useful when activities are regular and well defined, and requirements for data storage and security are limited; when accesses are frequent, continuous or unpredictable, there are time constraints to meet or data processing (internal searches, for instance) to perform, active tags may be preferred. Trolleyscan sell a Notetrack Management system. The client needs to supply the computer that will be used at the guard point. This system needs to have Win98SE/Win XP or Win Vista for an operating system. The administrator section of the software can be run on any existing computer. The package comprises of 1 UHF Trolley Scan fixed reader, 50 Ecochip tags, 1 Guard point software package and 1 administrator software package. Additional transponders and readers can be bought as needed. Price: £3000. A shopping centre however would need from 10-100 however. The system can be purchased from Trolley Scan at www.trolleyscan.com

Contact: Mike Marsh, info@trolleyscan.com

4. WiFi Tracking Solutions

802.11 Wi-Fi networks are available in most public buildings. The signals transmitted by the Access Points (APs) provide a readily available network of signals which may be used for positioning. The wide availability of existing Wi-Fi networks and of Wi-Fi enabled mobile devices makes WLAN positioning an attractive option due to the low roll-out and operational costs. The majority of systems in use today rely on measurements of RSS, Signal to Noise (SnR) ratio and Proximity Sensing. Each beacon (AP) sends out periodic broadcasts on the up or down link. Measurements are taken at the terminal device for RSS and SnR. Passive scanning is used to listen for the signals from the beacons. This is normally used to select the best signal for data communication. Each beacon emitted from an AP contains some information about the AP. For positioning purposes, one of the interesting properties is the Basic Service Set Identifier (BSSI) which acts like an individual name for the beacon. These beacons are emitted periodically and the time delay can be configured but is usually in the order of a few milliseconds. With the information gained from these beacons a number of positioning methods may be implemented. The AP with the strongest signal is considered to be the location of the mobile device. If the Base Station's (BS) coordinates are known to be (x, y), then with proximity sensing, the Mobile Device's coordinates are also considered to be (x, y).

WLAN fingerprinting is the most successfully used method in commercial systems available today. It is used in both the Ekahau system and the LA200 systems from Trapeze networks. There are two separate stages in the fingerprinting process, the offline and online stages. The offline stage involves calibrating the area where positioning is to be conducted. This can be a time consuming process and involves manually walking around a building with a Wi-Fi enabled device which is constantly taking "RSS snapshots" of the signals that it can detect at each location from all the detectable APs. This must be done every few meters or so and at each location a full 360 degree rotation must be carried out as there can be a large variation in RSS values depending on orientation. This information is then stored in a database with the coordinates of each location corresponding to a different pattern of RSS values. Systems such as Ekahau display areas where calibration has been conducted with their RSS values denoted by the different colours graphically on a map. The online phase involves actually getting a position fix from a mobile Wi-Fi device at an unknown location in the test area. A number of approaches can be followed with either terminal, network or terminal–assisted being used. The detected

RSS values at a particular location are compared with those in the database. The closest matching pattern with its corresponding location coordinates are given as the previously unknown coordinates of the mobile device. A number of different methods may be used for find which of these patterns is the closest as there will very rarely be an exact match. Disadvantages of this method include a time consuming calibration/training process. In addition, if some of the APs are moved then partial calibration needs to be redone.

4.1 Ekahau

The current market leader in Wi-Fi positioning systems is the Finnish Company, Ekahau[7]. The Ekahau Real Time Location System is a software suite that uses an existing WLAN network without the need for additional special network hardware to determine the location of a Wi-Fi equipped device. The suite has three main components namely the Ekahau Site Survey (ESS), the Ekahau Positioning Engine (EPE) and the Ekahau API that utilises the EPE system to create custom applications (see Figure 9). The EPE uses software based algorithms to calculate the position of a tag. However before the EPE can determine the location work it needs the site survey calibration information from the ESS. The ESS collects the information on the coverage and RSSI of each AP in the network across the area to be covered. The ESS gathers the calibration information by a person carrying the system and walking around the area to be covered. Ekahau provides an open API with support to integrate XML thus full visibility across geographically-dispersed campuses is available out of the box without the need to install software or hardware at remote sites.

Figure 9: Ekahau Components out of the box

The Client communicates with the mobile device's Wi-Fi chip and retrieves the RSSI information and passes this along to the EPE. The EPE is a positioning server that provides the location coordinates (x, y, and floor) of the mobile terminal or Wi-Fi tag (See Wi-Fi tags in Figure 10). The Ekahau manager merges information from the EPE and the Client and also provides applications for site calibration (Ekahau Site Survey) and live tracking.

---

[7] http://www.ekahau.com/

Figure 10: Ekahau Tags

A noteworthy element of the Ekahau systems is their proprietary "Rails" software which allows for tracking to be carried out in a way that replicates human movement and eliminates the "jumping through walls effect" that is common with other RTLS. This gives a unique competitive advantage to the Ekahau solution. The Rails are added by an administrator to "teach" the solution where devices are able to travel.

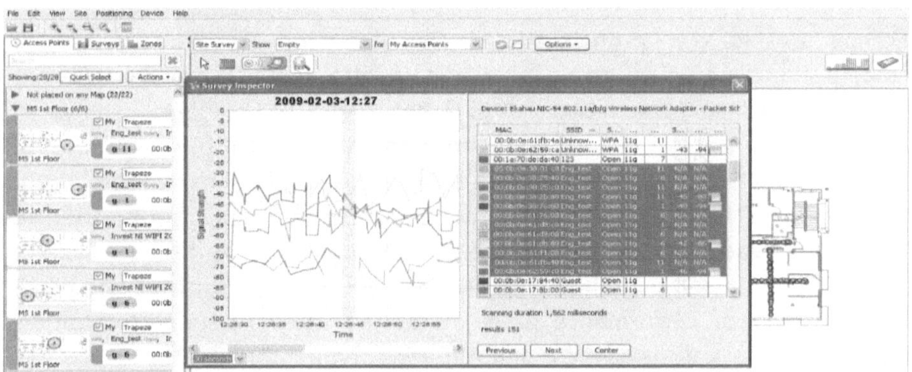

Figure 11: Ekahau Survey Inspector showing 30s fluctuations

The software views the area where the rails are as a higher probability of true location. Ekahau can use a network, terminal or terminal assisted approach. It also comes with an Application Programming Interface (API) to enable custom applications to be developed. The Ekahau RTLS system facilitates all tracking of devices as it does not rely on proprietary infrastructure or readers in order to track devices. The existing 802.11 Wi-Fi network is used for all tracking with signal strengths being recorded as they are. Ekahau Site survey records RSSI data of the test area with all observable aspects of the WLAN being considered. RF characteristics e.g. multipath, reflection, are recorded and do not harm location accuracy or signal measurement. This survey data then facilitates building tracking models. The observed client data is recorded and each recorded location is assigned a probability based on this data (see Figure 11).

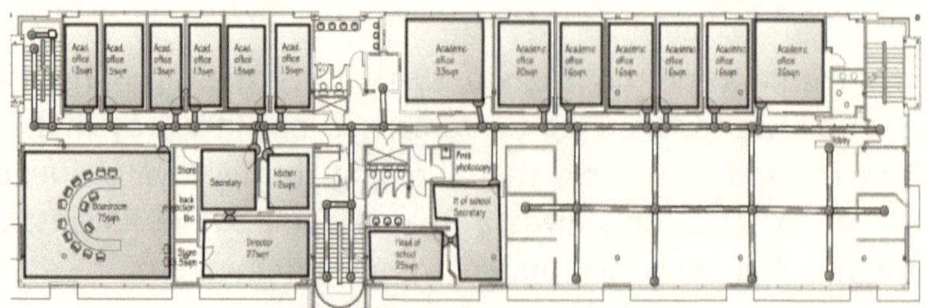

Figure 12: Rails and free space

Ekahau uses its own probabilistic location detection algorithms which are computationally efficient giving 1-3 metres accuracy in ~ 5 seconds. Different Wi-Fi devices "hear" the network at different levels (RSSI readings) even when they are located at the same distance from the Access Point (AP). A process of normalization is applied which allows for the use of hardware from different vendors. Figure 12 illustrates how the rails tools in Ekahau can be used to designate areas such as hallways between rooms. The Ekahau Planner provides real-time visualizations for displaying RF coverage shown in Figure 13 and a variety of performance parameters. The software supports multiple different wall material types and antenna types for the best possible calculation of RF signal propagation.

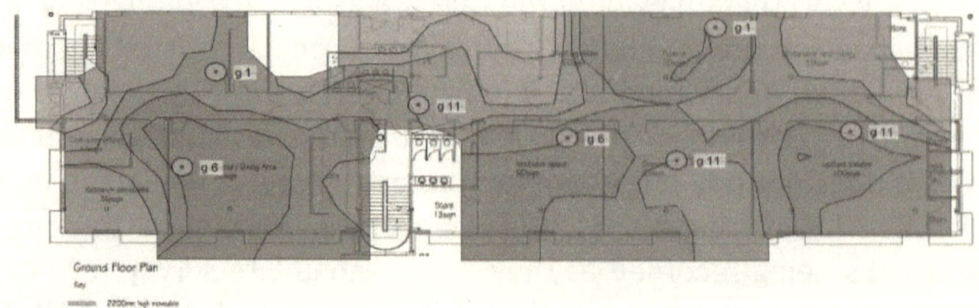

Figure 13: Shows RF coverage

The Ekahau planner presents an innovative approach that streamlines Wi-Fi network design and deployment (Ekahau, Planner data sheet). It can be used to intelligently simulate the initial access point placement settings, walls, and thus predict the expected network performance, prior to installing any Wi-Fi infrastructure (Ekahau, Planner data sheet). An easy to use drag-and-drop GUI is included for access point and wall placement on a facility floor map (Ekahau, Planner data sheet). The Ekahau Planner provides real-time visualizations for displaying RF coverage and a variety of performance parameters. The Ekahau

Software Development Kit (SDK) is an application that contains Java package, Javadoc, and code example for quickly connecting to the Positioning Engine. Companies such as Lever in the UK provide services to install such a system but they cannot provide any estimates of price until they visit a location and do the initial site survey.

Contact: Terry Bolton   Email: terryb@lever.co.uk
LEVER Technology Group PLC, Gelderd Point, Gelderd Road, Leeds, LS27 7JP
Tel: 0113 398 3300   Email: info@lever.co.uk

4.2 Trapeze Networks

The Trapeze Networks Location Appliance LA-200 is a rebranded Newbury Networks Location Appliance. Newbury Networks strategic business partners rebranded the unit under several different names namely the Meru Networks Meru RF Location Manager, the Nortel Networks Nortel WLE2340 and Trapeze Networks Trapeze LA-200 (See Figure 14). It will only be referred to as the LA-200 in this report.

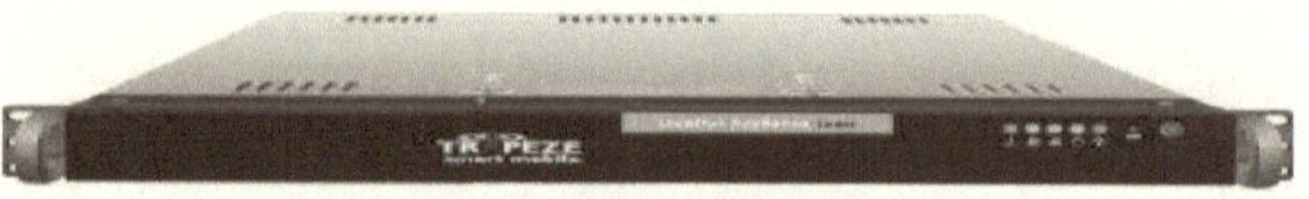

Figure 14: LA200

The LA200 uses server-side RSSI pattern matching techniques to locate devices or tags and claims the industry's highest performance for accuracy (i.e. claims to locate all devices to room level with accuracy at 99% with 10 meter precision in fewer than 30 seconds. It claims an ability to track up to 2,000 wireless devices without the need for specialized hardware or software on the tracked devices.

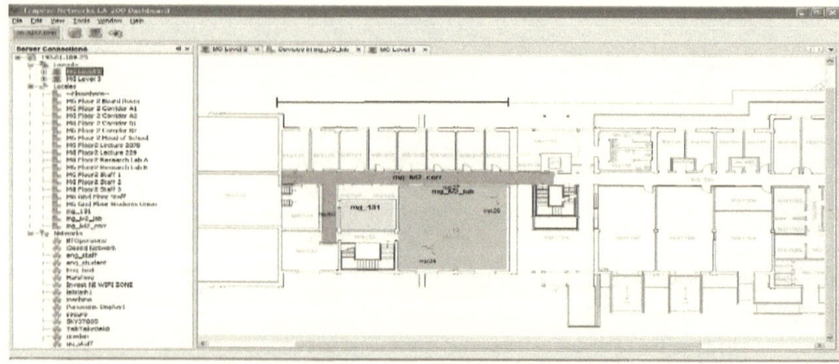

Figure 15: Trapeze Dashboard for the LA-200

The LA200 is shipped with a dashboard application (see Figure 15) which allows the viewing of real-time movement of WiFi devices, people, and asset tags on each floor. It comes also with some scripting examples for the API which enables custom applications and business-process integration with location services. The system is also able to store location history for each tracked device for up to 30 days. The Trapeze Networks Location Appliance provides the ability in real time to quickly and accurately locate and track assets, people or practically anything that is attached to an existing Wi-Fi network. It provides the capability to run custom or enterprise applications that require the ability to provide location sensitive content or security and track assets. The LA-200 is Wi-Fi compliant which allows the system to use the active Wi-Fi tags from other companies such as, AeroScout, Ekahau, Newbury Networks and Pango.

A summary of devices connected can be viewed through the Dashboard and options are available to view devices by server, locale or network as shown in Figure 16. An image file containing a scaled plan of a building can be imported into the system and different locales added by the user.

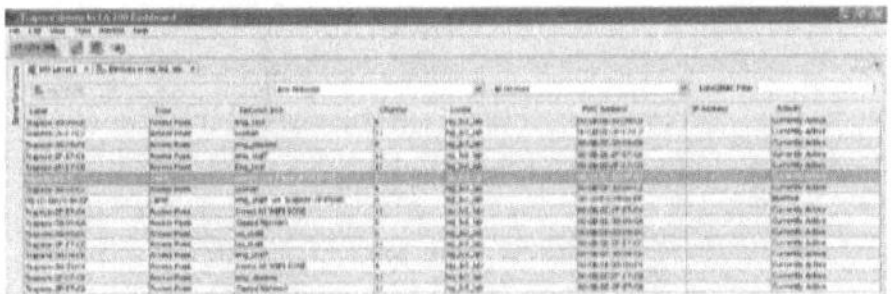

Figure 16: Trapeze Dashboard Device List Screen Appliance

The light blue area in Figure 17 represents the 'MG122 locale. After the locales have been identified the actual physical fingerprinting is done. The points where fingerprints have been taken are represented by a green flag as shown in Figure 17.

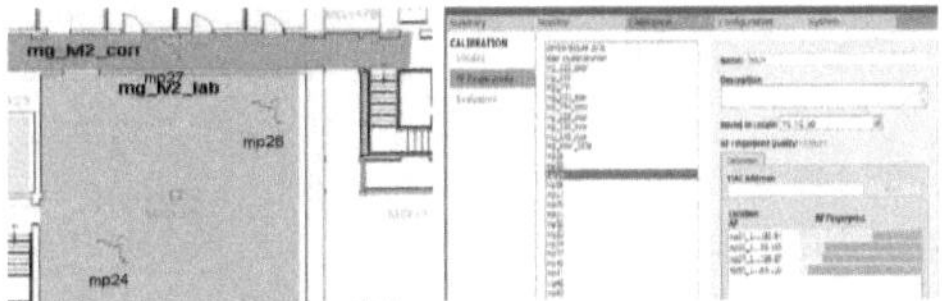

Figure 17: Dashboard fingerprint locations

Figure 18: LA200 Web Configuration screenshot

The Dashboard application summary screen displays information on the system, network and devices visible to the network. By clicking on a particular device from the device list screen the system will display the recent activity of the tag. A MAC address and date range for a given time period of up to thirty days can be entered here and the system will display the location of that tag for the given date range. The system will also display the fingerprint information it holds for a particular fingerprint point. For example in Figure 18 fingerprint 'mp24' is shown. Figure 18 shows a screenshot from the web based Dashboard fingerprint section which shows a fingerprint location in a graphical representation using green and purple bars.

Contact: Keith Bradley (Sales Engineer)
Trapeze Networks, 33 Clarendon Dock, Laganside, Belfast, Co Antrim, BT1 3BG
E kbradley@trpz.com   M 07515 660 846   Skype Trapeze_kbradley

4.3 AeroScout

Aeroscout is one of the leading manufacturers of RTLS systems. The company invented the first WiFi based RFID tag.   The company utilise standard WiFi networks to track the location, condition and status of assets both personnel and equipment.   The use of standard WiFi networks helps to reduce installation costs. Aeroscout has attracted many Fortune 500 companies with its use of standard WiFi system meaning a lower overall COO and a higher ROI.

As most shopping centres have a pre-existing wireless network on site and there should be no need to purchase additional hardware. Mounted exciters and choke points are used to localise the active tags. There are two types of exciters in the system, the mounted exciter (Figure 19) and the choke point exciter (Figure 21).   The mounted exciter will provide definitive location of people with the tag and provides the ability to make the LED in the tag (Figure 20) flash to determine its location in a heavily populated area. The choke point is set to scan for tags passing through every 200 milliseconds.

Figure 19 - Mounted Exciter

Figure 20 - Active RFID Tag

Figure 21 - Choke Point Exciter

The choke point enables large amounts of data to be acquired in a short space of time. It is particularly effective at doorways and in corridors. Tags that will be used with the system are Aeroscout TAG-2000. These tags have an outdoor range of 600m and indoors range of 180m. The tags operate on 802.11b/g @ 2.4ghz and have clear channel sensing to avoid interference with wireless networks.

The transmission interval can be set between 128msec to 3.5 hours. The tags have 3.6volts lithium ½ AA battery which is replaceable. The battery may last up to 4 years dependant on use. The mounted exciters are EX-3210. These exciters have a working range of 20cm to 3m. The power input is 12volts over POE (802.3af) hence the need for the POE switches. Choke point exciters are EX-2000B. These have a range of 50cm to 6m. They have an input of 24VDC

POE (802.3af), again, as with the EX3210 these must be powered by the POE switches.

Contact: Camworth, Avionics House, Kingsway Business Park, Quedgeley, Gloucestershire, GL2 2SN

T: 01452 881330   E: connect@camworth.com

4.4   Motorola Proximity Awareness and Analytics

Motorola Solutions' Proximity Awareness and Analytics software module allows enterprises to detect, analyze and act on location information from Wi-Fi devices. Their solution enables the identification of customer proximity and supports interaction using rule-based push-to-deliver benefits like personalized messages, coupons or assistance.

Their Analytical component provides customer insight with detailed statistics about customer activity (per store visit time, repeat customers, total customers in store, demographic profiles) and supporting programs to improve the customer experience It allows target zone definition which supports capabilities like tailored assistance or zone-specific offers and allows tracking of Wi-Fi device location and path over time. This supports optimization of an enterprise environment, a retail store for example, to minimize congestion and improve product placement. It comes with an open API allows sharing of analytics data with third-party applications, creating a vendor-agnostic environment for development and delivery of location-based services which add value, improve service and build loyalty.

The benefits are that the existing Wi-Fi infrastructure does not need replacing. The software based capability uses information from the existing WLAN network.

Contact: Graham Florence, Account Manager
Motorola Solutions UK Ltd, Government and Enterprise - Ireland & Scotland
Mobile: +44 7710 803346

4.5 Insiteo

Insiteo provide an indoor location solution for smartphones that leverages the Aruba Wi-Fi beacons for position information. The smartphones collect position information from the Wi-Fi network and forward it to the hosted server, which then calculates the user's position. The central server can interface with added-value applications, which are then pushed to the user. For example, customer retention can be enhanced by pushing coupons for services to patrons that approach exits, or patrons can be guided to vendors that pay for enhanced exposure.

Insiteo also delivers indoor geo-location services for smartphones. The solution combines indoor position location via Wi-Fi with a mobile application to locate the user's position and nearby points of interest – or directions to those locations – on a map. Target markets include large public venues such as shopping centres, airports, metro and train stations, convention centres, museums, and hotels.

Contact: Augustin Monsaingeon   T: 33 6 23 11 71 91
http://www.insiteo.com

5. Bluetooth Tracking Solutions

There are a number of new technologies available that use both Bluetooth to augment Local Positioning Systems. Most Bluetooth systems similar to WiFi and active RFID systems are made up of three major components the positioning server, the access points and the tags. The tags can be either special vendor specific Bluetooth tags or any Bluetooth equipped device such as a mobile phone. Some systems claim up to ninety five percent reliability accuracy to around two meters. Bluetooth tags are detected by several methods namely, using RSSI to triangulate the location, putting an access point in every room and using the nearest AP to the tag to indicate its location.

5.1 BlipTrack

BLIP Systems offer a range of solutions for tracking & mobile marketing. The BlipTrack solution brings real-time information about movement to achieve transparency, improved documentation and efficiency of daily management operations. They have installed it in a number of European airports. The Blip System, based on the Bluetooth radio technology, is able to track mobile units (e.g., phones) that have Bluetooth and are set to visible mode. It performs tracking by a number of Blip nodes, each of which is part of a *Blip zone*; whenever a Blip node detects the radio signal from a Bluetooth unit appearing or disappearing, it sends a signal to applications using the Blip System. Applications can then provide location aware services, based on the zone information thus captured. A Blip zone is the area in which a mobile unit will be detected by a Blip node in this zone. The size of a Blip zone is determined by the signal strength with which the Blip nodes detect the mobile units, and can be changed by adjusting it using a system administration tool.

Bluetooth addresses and other metadata. Bluetooth devices are very common in mobile phones, PDAs and laptops. Bluetooth transceivers are available in three classes, defining the maximum power output and thereby range of the signal. A class 1 transceiver has the highest power output and with it the longest range of up to 100 meters. Class 2 and 3 devices have a range of up to 10 meters and 1 meter, respectively. The range of class 2 and 3 devices can be slightly extended due to higher sensitivity by using class 1 access points.

Most mobile phones come with Bluetooth modules, meaning that the best case penetration is nearly as high as for TMSI (used by Path Intelligence), but because they have to be activated to be seen, there is a moderate penetration rate when using this technology for tracking. BLIPs system queries for unique devices once per second. The Bluetooth system provided by BLIP is ready for communicating location and context specific content on the network.

The range of a typical Bluetooth tracking system however is only around 20-30 meters. The short range requires many access points to be installed in order to cover the area. If the system should be upgraded to support triangulation, many more access points would be required. Due to the fact that most Bluetooth transceivers in mobile devices are class 2 transceivers and the signal output of different vendor Bluetooth modules varies greatly, their range is limited to approximately 10-30 meters with a precision of about 5-10 meters.

Contact: BLIP Systems A/S Hækken 2, Vester Hassing, DK-9310 Vodskov, Denmark
P   +45 9825 8200   E   sales@blipsystems.com   W
http://www.blipsystems.com

5.2 Direct enquiries - Indoor Tracking System

Direct Enquiries offer an Indoor Tracking and Navigation system to view and record movements of people and assets throughout a building. The system works by networking special Beacons throughout a building, which pick up signals transmitted via Bluetooth Low Energy from our Tracking Tags. The Tags can be placed on anything from staff to property. The data is then analysed and a business can view both live and archived data. Plans of the venue are digitised and held in their web servers. When accessed via secure passwords, Tagged assets and personnel can be seen moving around the venue, with accuracy of between 0.5m-1m. The images can be viewed live or archived to view at a later date for security or SLA purposes.

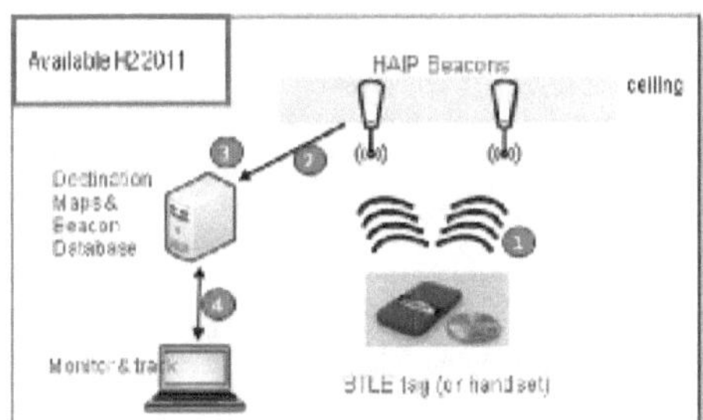

Figure 22: Indoor Tracking Architecture and screenshot

They claim their solution can allow organisations to track personnel in real time, within 1m accuracy indoors, receive evidence based SLA reporting and of course track people around buildings.

They also offer integration with some smart phones, which allows viewing movements on handheld equipment. This also allows buildings to be managed from anywhere in the country.

Contact: Greg Hewett, Senior Business Development Manager, Direct Enquiries LTD, Amber House, Market Street, Berkshire, RG12 1JB

P 01344 360 101   M 0774 8980519

http://www.directenquiriesltd.com/tabcontent.aspx?service=indoor-tracking

6. Mobile Tracking Solutions

In a mobile phone system the position is given as that of the cell that the phone is currently in.   This is known as "cell_id" based positioning. It can be network based or terminal based.   Some towers broadcast their location intermittently.   This "beacon" can be listened to and the location deduced from it. This approach has been used in a number of indoor systems. There are also GSM IMSI 'catcher' products out there. These allow the installation of 'rogue GSM' towers to collect basic GSM information about hand-sets in the local area. One company I spoke to called NeoSoft based in Switzerland sell a 850/900/1800 or 1900 MHz model transmitting at 10Watts for £40,000 and their
3G 2100 MHz catcher transmitting at 5 Watts retails for £60,000. One company that has garnered most press interest is Path Intelligence and their equipment is more accessible than that of NeoSoft (and less intrusive).

6.1 Path Intelligence

FootPath from Path Intelligence is the UK's first commercially available example of a reality mining tool. That is, it uses machine sensed data to understand human behavioural patterns. FootPath works by detecting anonymous transmissions from mobile phones carried by centre shoppers. The FootPath system consists of a small number of discreet detector units installed throughout the centre. These units calculate the movement of consumers without requiring the shopper to wear or carry any special equipment. The units measure signals from the consumers' mobile phones using unique technology that can locate a consumer's position to within a few metres. The units feed this data (24 hours a day, 7 days a week) to a processing centre where the data is audited and sophisticated statistical analysis is applied to create continuously updated information on the flow of shoppers through the centre. At anytime the shopping centre management can access the data via PI's secure web-based reporting system – PI Explorer.

It works by listening in on the network communication between the mobile phones and base stations using strategically placed monitoring units. They use an identification number contained in the communication packages called the Temporary Mobile Subscriber Identity (TMSI) to isolate and track the individual phones within the area. TMSI is an ID assigned to a cell phone when it enters a new location (base station) on the GSM and UMTS network. The TMSI is periodically transmitted between the device and the base station and is also used when creating communication channels for receiving or making calls. The ID is temporary and changes over time and location. Path Intelligence Ltd. has developed a method allowing them to track the devices and present results from data analysis, such as shopper flow and densities in a mall, to the customer. As the system is able to track any mobile phone as long as it is turned on, the penetration rate is very high. The range of a mobile phone is rather large up to 3-20 miles depending on the amount of interference and obstructions in the area, therefore the number of nodes needed to monitor signals in the airport can be kept at a minimum depending on the accuracy wanted. The system provided by Path Intelligence has an accuracy of down to 1-2 meters using triangulation.

However, because the TMSI is only broadcast periodically, e.g. when the mobile phone is being used or changes base station due to low signal strength, the time between updates varies from a couple of seconds up to a number of minutes. Additionally, because the TMSI is changing continuously, there is a possibility that the tracking sessions end prematurely, thereby obscuring the data collected. The system provided by Path Intelligence provides no means of communicating to the tracked devices. Because the system relies on triangulation for pinpointing the position of a device the complexity of the tracking system greatly increases. Path Intelligence charge by the square footage for installation of their systems.

Path Intelligence are currently testing in a few select locations in the UK their latest product called *Flockr*. Flockr will allow tracked individuals to be identified. Their mantra for this product is "Detect, Analyse and Influence" http://www.goflockr.com is registered to Path Intelligence.

Contact: Path Intelligence, 1000 Lakeside, North Harbour, Portsmouth, Hampshire, PO6 3EN

Personal Contact: Joe Sullivan, Vice President, Business Development
E: joe.sullivan@pathintel.com
Mobile: 203 512-9432   Office: 212 537=9252   Skype: js-pathintel
London Office : 0844 587 0801   E: info@pathintelligence.com

6.2 Nearby Systems

Nearbuy claim to track location within one meter using data from infrastructure that retailers already have in place. There are no sensors, software or beacons to install.   Their appliance is the only addition. Headquarters store managers can access micro-location analytics through role-based dashboards. Nearbuy's analytics links in-store activities with guest WiFi analytics, in real-time. Nearbuy micro-location data claims to provide store leadership with visibility into shopping patterns at each store and help managers use sales associates more efficiently by tracking the location of employees equipped with mobile devices. The system uses signal strength measurements from APs and floor plan data to track location and video surveillance system as a "presence" detector to help their LocalEyes algorithms calculate best possible location of any shopper. It operates with Motorola Solutions and Aruba Networks infrastructure and can connect location to identity with opt-in information captured by the Nearbuy Captive Portal.

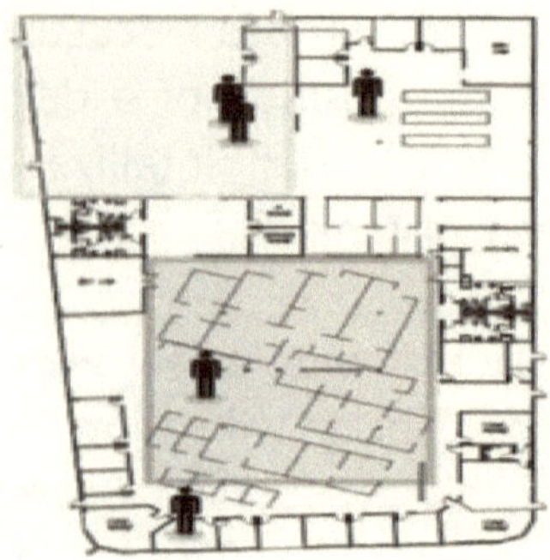

Figure 23: A screen shot from a Nearbuy Systems demo video

Nearbuy LocalEyes is a managed appliance deployed in individual stores. Nearbuy Luneta, running in the Nearbuy data center, serves as the primary, centralized management interface for LocalEyes. This deployment approach delivers optimal performance for the retailer's network while making it cost-effective to maintain the historical data needed for ongoing trend analysis. LocalEyes runs on a standard PC-based server running the Linux operating system and uses an Nvidia graphics card for video processing. LocalEyes is typically placed in a store wiring closet and needs LAN access to the store wireless LAN and video systems. Integration with the video system can be directly from IP based cameras or by integrating LocalEyes with a DVR for CCTV based systems. LocalEyes and Luneta need to be able to communicate with each other over a WAN connection.

More Info: http://www.nearbuysystems.com/products/micro-location.html

6.3 LocationLabs

LocationLabs[8] offer a number of location services. They are US based and predominately focused on the US market. They have plans to roll out in the UK soon. They offer a Universal Location Service (ULS) which gets you the location of any mobile phone across carrier networks, with no download, via a simple cloud API. They claim they can remotely locate 300MM+ mobile phones. They also offer a geofencing product. At its core, the Geofencing product is a client SDK with background processing that enables creation of a geofence, a virtual perimeter around a location of interest, and triggers an alert when an application user enters or exits this perimeter.

The product comes with a downloadable client SDK and a server component that provides web service functionality, scale and analytics. Developers can create geofences from within their application, or use POI locations sourced from 3rd party providers made available by Location Labs. Developers can also leverage pre-defined actions such as Auto-Checkin and Checkout, time sensitive notifications, or create custom actions based on entry, dwell and exit events.

When used in conjunction with the spatial storage product, Geofencing provides application ability to identify users, devices and events within a geofence in recent past helping drive useful application-specific actions. In addition to all the above, the client SDK minimizes device battery drain by intelligent use of GPS and device network OS level functionality to detect location when running in the background. Geofencing is currently available on iOS and Android platforms.

---

[8] http://www.locationlabs.com

6.4 GloPos

GloPos is a developer of a software-only positioning technology that makes all mobile phones location aware - outdoors, indoors, and even underground. It has confirmed an indoor positioning accuracy of 7.7 to 12.5 meters in an independent test of its software conducted by VTT, The Technical Research Center of Finland.

The stationary indoor and underground tests show that GloPos is capable of non-filtered average positioning accuracy of 15.1-23.9 meters. 75% of the measurements present an average position accuracy of 7.7-12.5 meters. With filtering, the indoor accuracy of GloPos technology is as good as the outdoor accuracy in an urban environment with a normal GPS embedded in mobile phones. When comparing with state-of-the-art cell positioning (Google Maps) they claim the average accuracy is at least two times better and in some cases even ten times better (indoors).

Mikael Vainio, CEO
M +1 415 349 5966 or +971 50 667 0349
E mikael@glopos.com
W http://www.glopos.com

7.   Indoor Map Based Systems

A number of indoor solutions based on cell-tower triangulation or Wi-Fi network databases have appeared in recent years, the newest being one from Point Inside. By combining a proprietary location solution with indoor maps of major malls and airports, these systems offer guidance in places where Google Maps and others simply cannot. It is a rapidly growing area of localisation. Google, Yahoo, Microsoft, Nokia and a few standalone portable navigation device makers are well-entrenched with their efforts to dominate the indoor navigation space. An overview is provided here of the recent offerings in creating indoor maps for large public spaces.

Some of these systems use existing WiFi infrastructures to triangulate position. The key difference between these systems and the ones outlined elsewhere such as Ekahau and Trapeze is that you are reliant on the company or other users in having mapped the location beforehand. They are however very powerful and cheaper to utilise although there will be licensing costs to build your own apps with integrate with their API. Google and Apple are both in the positioning game and have both recently come into the news for collecting our personal data to improve their location abilities. They are gathering location information to build massive databases capable of pinpointing people's locations via their mobile phones. These databases could help them tap the £2 billion market for location-based services, which is expected to rise to £6 billion by 2014 according to Gartner. Generally, when companies have collected data from users it has been from personal computers. The data gathered through this medium can be tied only to a city or area code. The rise of internet-enabled mobile phones allows the collection of user data that is much more personal and can be tied to locations on a more granular scale.

7.1 Point Inside

Point Inside provide maps of indoor destinations for use with its mobile smartphone indoor positioning applications. It has just released an IPhone product. It is partnering with Meijer stores for a pilot program that will enable shoppers to "see the location of more than 100,000 items in a retail supercenter using their smartphones." "Point Inside's platform facilitates the aggregation of destination-specific content, adding the critical context of *location* to each item. This marrying of *'what'* to *'where'* enables shoppers to quickly find what they need at their destinations, whether it's a specific department, service, like a fitting room or restroom, or even an individual product on the shelf."

If a location has been mapped by its software then it provides directions on a smartphone to easily navigate say from the GameStop to Lenscrafters. Point Inside says it now has maps covering over 100,000 stores, gates, kiosks, restrooms, elevators, and escalators in U.S. and Canadian malls and airports. And much like Foursquare, it offers retailers a geo-based way to provide promotions to a nearby consumer audience. Point Inside is actively soliciting for venues to use the Point Inside "Indoor Smart Map" technology. Malls and airports are just the lowest-hanging fruit for Point Inside. If the company can build up a large enough base of venue directories to mash up with its location solution — say, large office properties or sports venues — it could own the indoor navigation space.

Point Inside is attempting to transform the shopping experience by enabling retailers to engage proactively with customers through their smartphones at every point along the purchase path. Point Inside's unique, patent-pending micro-location and indoor mapping technologies expand and enhance the value of smartphone apps by bringing location-based services into the store, letting retailers know which section and aisle their customers are in. Combined with understanding of the customers' purchase intents from the shopping list, purchase history and Point Inside's engagement technology, retailers can now connect customers with highly relevant messages while they are inside the store shopping. Combined with understanding of the customers' purchase intents from the shopping list, purchase history and Point Inside's engagement technology, retailers can now connect customers with relevant messages while they are inside shopping. This technology does allow tracking of people.

Contact: http://www.pointinside.com/

7.2 Google Indoor Mapping

Google also wants to map notable indoor locations like airports and retail stores to provide floor by floor navigation to users. It has just released a new version of Maps for Android that is fine tuned for indoor navigation. Google Maps is launching indoor navigation with floor plans for a few dozen airports and retail locations in the US and Japan. The list of supported locations in the States include Mall of America, IKEA, The Home Depot, select Macy's and Bloomingdale's.

The newly added indoor maps do not offer turn-by-turn navigation but the provided layouts should help usher visitors along to the nearest bathroom, clothing shop or elevator. All of the positioning information is culled from the same set of data (including GPS) used for "My Location,". It has even been optimized to detect movement along the z-axis so that with a feature called "Automatic Floor Detection" - it can keep track of your progress as you move about from escalator to escalator. Google's is also endeavouring to extend its indoor reach, opening up its mapping inventory with a self-service tool that will allow business owners to upload floor plans directly to Maps. Business owners to manually add floor plans to its map database at maps.google.com/floorplans. A full list of buildings with Indoor Maps coverage is available at Google's Help Center.[9]

---

[9] http://www.google.com/support/gmm/bin/answer.py?hl=en&answer=1685827&topic=1685871

Google has just recently launched a crowd-sourcing Android app, the Google Maps Floorplan Marker, to allow users (i.e., business owners) to provide feedback about how accurate the Google Indoor Location service is for their venue. The app guides the user where to go inside the venue and do some WiFi scanning (and even Cell ID sniffing). This process collects the necessary data that Google needs to improve its indoor location service. Google is hoping these business users will help Google calibrate the Z-level (floor level) positioning challenge.

7.3 Wifarer

Wifarer is another company offering an indoor positioning technology app to steer users through large venues such as malls, airports, convention centers and museums. The free app, available for iPhone and Android users, has an interactive location directory to help users find a specific store in the mall or terminal at the airport. Wifarer claim to pinpoint the user's location within 4-and-a-half feet, and draws a path to the location the users wants to go. The big hook here is the app will also show promotions and sales based on a user's location — providing advertisers a new way to catch an audience. Individual stores in a mall can control the ads customers are viewing. Once a customers steps in the store, the brand has complete control over what the app user sees. The store could provide inventory and directions around the store if they wanted. Most stores will likely use this control to target customers with advertising specific to their store. Interestingly, Wifarer is creating its indoor positioning systems by using each venue's pre-existing Wi-Fi, and then creating radio frequency fingerprints to infer users' locations. All locations at present are in the US and Canada.

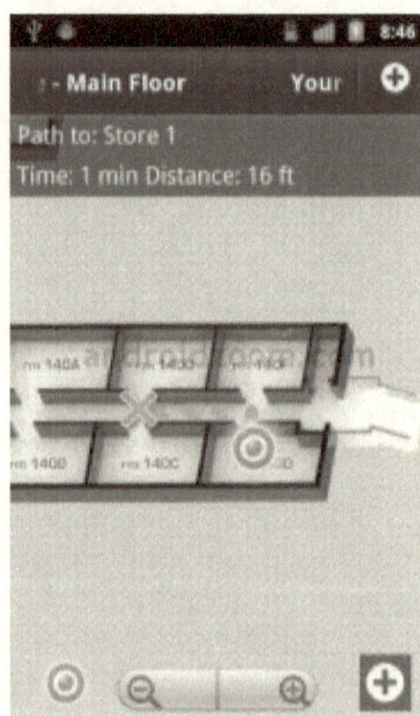

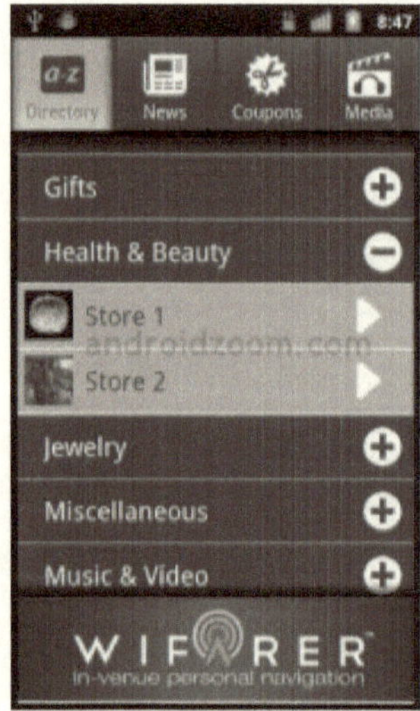

Figure 25: Wifarer screens

Wifarer's Indoor Positioning System still not disclosed but it seems to add features from the phone to supplement standard WiFi positioning. Otherwise it could not claim to outperform current positioning technologies by refining a location accuracy to an average of 1.3 metres.

This works by firstly downloading the app at the Android Store and installing it on a smartphone. Next you need to get a floor plan of the location. Wifarer will then pinpoint a visitors location on the smartphone screen map. It automatically updates location on the map as they move. Its online management system allow you to edit and update information in real time, which also appears on visitors smartphones instantly. Wifarer provides real time venue analytics. It anonymously gather visitors' smartphone data to produce heat maps and analytics. It offers insights into visitors behaviour in terms of where they go, how long they stay as well as their search & page view data regarding products. Users just need to download it once for free and can use it in any Wifarer certified venues in the world.

Contact: http://www.wifarer.com/

7.4 Qubulus

Qubulus provides a positioning platform for smartphones where GPS does not work. Application areas include navigation, point-of-sale applications and enterprise solutions. Location data is stored on the backend to create analytics for the physical space. The system attempts to provide visitors with exact positions and directions and at the same time can provide selected information about points of particular interest from exhibitors or vendors that the visitor walks by. Qubulus have also launched an API that will allow third-party developers to add indoor positional capabilities to their apps with 'shelf level accuracy'. This means that it will possible be able to add, for example, navigation to specific products in a store. The approach for developers is to overlay a to-scale floor plan of a building over Google Earth and get the GPS coordinates of each corner. The floor plan and coordinates are then imported into a special recording app, which is taken to the site and used to record radio signals at various positions. Once all the data is collated it is exported and sent back the Qubulus guys who generate a fingerprint file from it. With this file imported back into the recording app the phone will then be able interpolate GPS coordinates from the readings and show its position on the floor plan.

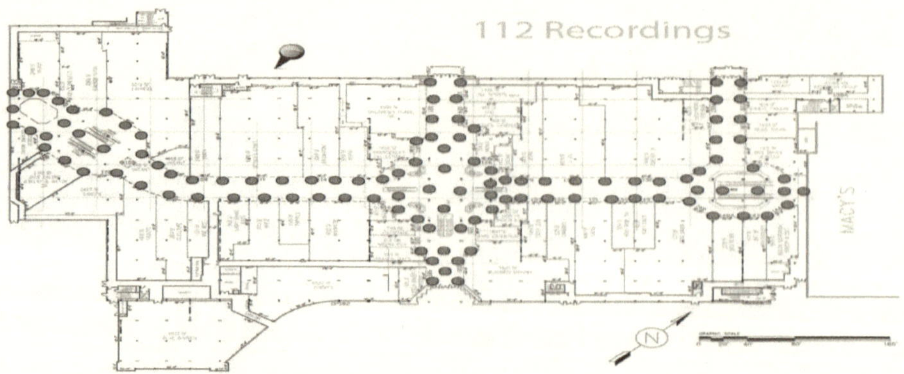

Figure 26: Data visualization of recordings performed at a mall in California

Qubulus is one of a number of companies working in the rapidly developing field of indoor positioning – GPS for inside buildings. At the moment the beta testing is closed. however when it was opened, one has to download the LocLizard API working in order to allow each specific location, to be mapped out with the Qubulus' Gecko service. This uses signals from radio networks to map out a building, allowing apps that use the data to locate users . Once you build up a grid of those spots and link it to a map this creates positioning which is accurate to three meters. Demos that I have seen are pretty impressive[10]. It could potentially be better than anything currently on the market.

What sets QPS apart from the competition is that its technology not only works on a horizontal axis but a vertical one as well. This means that you will know what part of the building you are in and what floor you are on, since the vertical positioning is accurate up to one meter. Current methods use triangulation and are accurate to 10-20 meters, which is not helpful, since there could be many walls between you and your destination. Qubulus currently has two pilot projects going with carriers in the US and UK.

Contact: Qubulus AB, Brogatan 7, 211-44 Sweden
P +46 (0) 70 575 33 12   http://www.qubulus.com

---

[10] http://indoor-positioning.co.uk/2012/03/achieving_indoor_positioning_with_gecko_by_qubulus/

7.5 SenionLab

SenionLab provide a similar indoor positioning and navigation technology to Qubulus. It seeks to improve navigation capabilities in environments where GPS systems are unavailable. It is intended to run on mobile phone platforms relying on sensors available in the phone. It is a cost efficient and easy to use, since it does not require that the buildings are equipped with infrastructure to determine the user's position.

NavIndoors for iPhone is a pedestrian indoor navigation system for indoor environments such as shopping malls (see Figure 27). NavIndoors for iPhone is delivered as a separate software module together with a fully documented Application Programming Interface (API) that can be integrated into your location-based iPhone, iPad or iTouch application. It is fully compatible with iPhone 3GS, iPhone 4, iPad, iPad2 and iTouch. Navigation and positioning is based only on the sensors available on the mobile device. No additional hardware is required besides the mobile device for navigation and positioning.

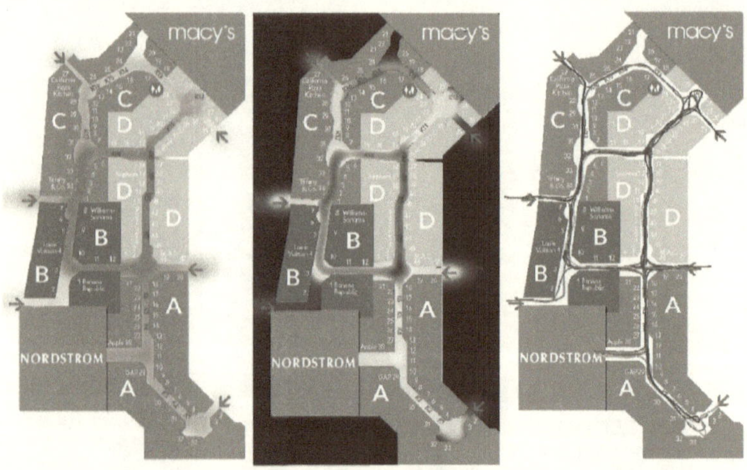

Figure 27: SenionLab Maps

NavIndoors for Android is also available with enhanced WiFi signal-based positioning capabilities for Android mobile phones. It is fully compatible with most of the Android based smart phones and comes with a documented Application Programming Interface (API) for easy integration with third-party location based application.

Both apps allow users to be monitored in real time and user behavior related charts can be automatically extracted. The aggregation of shopper paths for a specified period of time produces a "PathDensity" map identifying relative shopper traffic throughout the store.

Contact: SenionLab, Linköping, 583 33, SWEDEN
T   +46 703 496758   E   sales@senionlab.com
http://www.senionlab.com

7.7 Walkbase

Walkbase have offices in Finland and the US. They allow developers to take measurements inside a building via the built-in WiFi chip on Android devices and upload the data to their servers. Then via an API call, the app will compare its position to an online database of points of interest (POIs), which could be existing services like Foursquare or Factual or a proprietary database, and report back when it identifies that the device is in close proximity to one of the POIs. Rather than returning actual GPS coordinates, it returns a contextual message such as "inside the Wesfield Mall, Nordstroms, 3rd Floor". It is difficult however to imagine how you would then display that information on a map. It is also Android only however the API is free to sign up for.

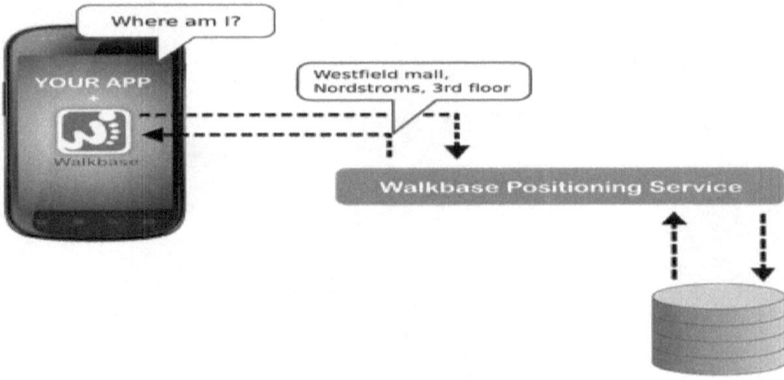

Figure 28: Walkbase

Walkbase offers a library that you can integrate with your mobile code to add indoor positioning capabilities for an Android application. The library provides developers with access to much improved indoor positioning that drives local room-level context as compared to what is possible with GPS only techniques. The library makes use of the built-in Wi-Fi radio chip most modern smartphones have. Measurements are performed using the chip and the obtained results are then sent to our servers. Instead of a physical location with longitude and latitude, the Walkbase Positioning Engine gives you a logical location (e.g. Starbucks, Radioshack) which is much more relevant and meaningful for most LBS apps. Once you have deployed an application with the Walkbase library embedded, you will immediately be able to see on a map where people are using your application. This will allow you to detect popular locations which in turn can help you refine the context. They say they are continuously developing analytics products in collaboration with the developer community.

Walkbase allows registered developers to upload and maintain their proprietary location database at their servers. This ensures your location data works smoothly with the Walkbase library. In most use cases, it is their recommendation to use existing POI services like Foursquare or Factual. They, however, offer this option to support use cases where refined proprietary POI database is needed. A mobile application can still access proprietary POI database through their REST-interface. Walkbase Positioning Engine is free for non-funded start-ups and non-profits. Email is required to register. The service is free to try for commercial use as long as the total number of API calls stays under 50k/month. Once the total amount of calls per month exceeds 50.000, you will move to a monthly plan starting at $40 for up to 200K/month.

Contact: Walkbase Ltd., Tykistökatu 4 D, 20520 Turku, Finland
E   support@walkbase.com   http://www.walkbase.com

7.8 Indoor Atlas

IndoorAtlas provide a system that utilizes the anomalies of ambient magnetic fields for indoor positioning. IndoorAtlas offers a complete software toolbox for adding and managing floor plans, collecting data to create magnetic field maps, and an API to use IndoorAtlas' location service for mobile applications. IndoorAtlas' core technology is independent of external hardware infrastructures (such as radio access points) and is able to pinpoint the location inside a building within 0.1 - 2.0 meters.

IndoorAtlas' cloud-based location service is illustrated in Figure 29. The application uses the IndoorAtlas API to communicate with the location service. The API sends processed sensor data to the location service, which computes the current location estimate and delivers the estimate back to the application's event listener method through the API. The location service connects to the map database, which hosts the magnetic field data collected from the building using the IndoorAtlas Map Creator application. The IndoorAtlas location service has been built on the top of Microsoft's Windows Azure cloud platform. IndoorAtlas' core technology is a software-only location system that requires, from the hardware point of view, only a smartphone with built-in sensors (no external hardware infrastructures, such as radio access points, are needed). They claim the accuracy in IndoorAtlas' technology in modern buildings ranges from 0.1 meter to 2 meters.

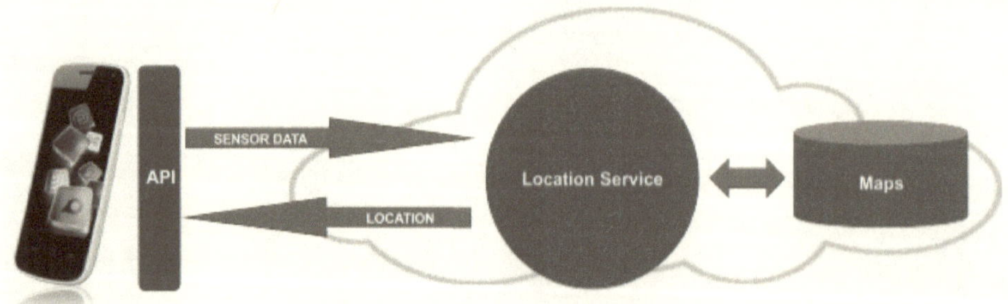

Figure 29: Indoor Atlas Cloud Location Service

Before using the IndoorAtlas' location technology, a magnetic field map must be generated from the part of the building where the location service is going to be used. IndoorAtlas offers a complete software solution in adding and managing floor plans, collecting magnetic field data and using the location service API. The process of creating location awareness inside a building starts by adding a floor plan image to IndoorAtlas Maps. The Floor Plans web application is used then to   align   floor   plans   with corresponding geographic coordinates, enabling the use of the geographic coordinate system in application software. After opening the floor plan in MapCreator, the user marks the planned route (typically a straight line or a curved path) on the smartphone screen. The user walks along the path and records the magnetic field data. The Map Creator application connects with IndoorAtlas Maps, which generates the magnetic field map that will be used for indoor positioning. Next, a location aware application can begin using the IndoorAtlas API for accurate positioning.

Contact: IndoorAtlas Ltd, Teknologiantie 14 C, 90590 Oulu, Finland
E info@indooratlas.com   W http://www.indooratlas.com

8. Miscellaneous Tracking Solutions

This section describes some other widely used indoor or local area positioning systems. Each of these has their own advantages and disadvantages and while in widespread use, they cannot solve all problems.

8.1 Ultra-wideband

Ubisense are a UK Company and one of the first to exploit Ultra-Wide Band for Real Time Location System (RTLS). Ultra wideband is precisely timed short bursts of RF energy to provide accurate triangulation of the position of the transmitting tag. Ultra-wideband (UWB) is a radio technology which can be used at very low power levels for short-range high-bandwidth communications (>500 MHz) by using a large portion of the radio spectrum. UWB transmissions send information by generating radio energy at specific time instants and occupying large bandwidth thus enabling a pulse-position or time-modulation. The information can also be modulated on UWB pulses by encoding the polarity of the pulse, its amplitude, and/or by using orthogonal pulses. Unlike conventional RFID systems, which operate on single bands of the radio spectrum, UWB transmits a signal over multiple bands simultaneously, from 3.1 GHz to 10.6 GHz.

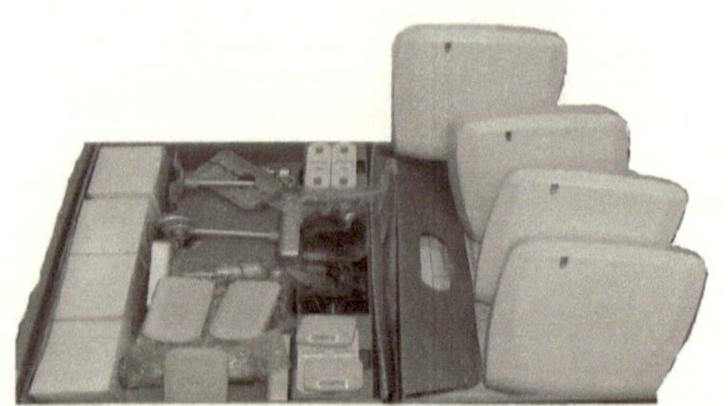

Figure 30: Ubisense out of the box contents

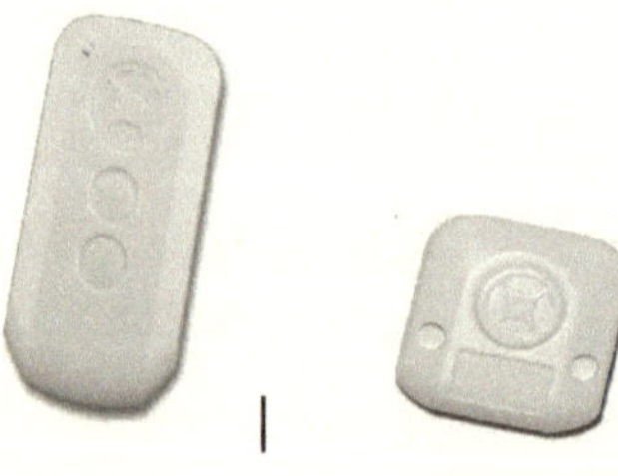

Figure 31: Slim & Compaq tags

In a UWB location system, small active tags are attached to the objects to be located, or are carried by personnel (Figure 30). The signals emitted by these tags are detected by a network of receivers surrounding the area. By detecting the signal at two or more receivers, the 3D position of the tag can be found. It is worth noting that two algorithms are employed. One calculates the time difference of arrival of a signal at two different readers and the other calculates the angle of arrival of the signal. The Ubisense RTLS solution utilizes battery-operated radio tags and a cellular locating system to detect the presence and location of the tags.

The Ubisense Series 7000 sensor is a precision measuring instrument containing an array of antennas and ultra-wideband (UWB) radio receivers. The sensors calculate the location of the tags based on reception of the detected UWB signals transmitted from Ubitags. Each sensor independently determines both the azimuth and elevation Angle of Arrival (AOA) of the UWB signal, providing a bearing to each tag. The Time Difference of Arrival (TDOA) information is determined between pairs of sensors connected with a timing cable. Sensors are administered remotely using standard Ethernet protocols for their communication and configuration. They work in standard wired and wireless environments, using networking infrastructures, such as 802.11 access points, Ethernet switches and CAT5 structured network cabling for communication between sensors and servers.

The locating system is usually deployed as a matrix of sensors that are installed at a spacing of anywhere from 50 to 1000 feet depending on the site layout. These sensors determine the locations of the radio tags. Ubisense consists of *Tags* - designed to be mounted on assets or to be worn by a person; *Location Engine* software to install and tune a Ubisense sensor network and track tags in real time, through a series of configuration wizards and the *Location Platform* software which provides persistent storage and distribution of real-time location events for multiple clients in conjunction with real-time monitoring and notification of user-specified spatial interactions between objects. Ubisense claim that their systems (see Figure 29) require fewer readers than other systems which implement only the time-difference algorithm. Readers receive data from the tags (max distance 160 m) and send location updates through the Ubisense Smart Space software platform. Ubisense creates sensor cells each requiring a minimum of four sensors. It claims to achieve scalability to 1000's of sensors using low-cost off-the-shelf servers over Ethernet. The standard range covered by a cell is less than <160m and typically the range between a tag and a receiver is 10m-30m. Key limiting factors include the level of building obstruction between the two. Ubisense claim an achievable range accuracy of <15cm even within a complex indoor environment. Accuracy such as this would allow location aware applications to pinpoint particular devices being used in a room.

Ultra wide band systems work well indoors as the short bursts of radio pulses emitted from UWB tags are easier to filter from multipath reflections than conventional RF signals, however metallic and liquid materials still cause some signal interference. Ubisense however claim that this can be overcome through the strategic placement of sensors and UWB also possesses the ability to determine "time of flight" of the direct path of the radio transmission between the transmitter and receiver at various frequencies which helps overcome multipath propagation. UWB pulses are very short in space (less than 60 cm for a 500 MHz wide pulse and less than 23 cm for a 1.3 GHz bandwidth pulse) therefore most signal reflections do not overlap the original pulse, and thus the traditional multipath fading of narrow band signals does not exist.

Ubisense uses active tags named Ubitags which are manufactured by C-MAC and designed to be easily mounted on the side of vehicles and assets or worn by a person (Figure 30). Their update Rate is 0.01Hz – 20Hz. These tags have a unique 32-bit identifier and broadcast a beacon including their location as often as 10 times a second (essential for tracking people walking quickly through a monitored area) or as infrequently as once every few minutes. This rate can be changed dynamically over a wireless link while the system is running and in response to individual tag behaviour. If a tag is moving quickly, the update rate of its beacon can be programmed to increase, and if a tag is stationary, the update rate can programmed to decrease. This feature is an attempt to conserve battery life. Ubitags are designed to last for approximately five years in typical use. The Ubitags can store up to 200 bytes of data, four of which are used to store the ID. Tag data can be changed over the network and individual tags can be paged. The Compact Tag is a small, rugged dust & water resistant device specifically designed for use in harsh industrial environments.

8.2 Ultrasound Positioning

Ultrasound signals may be used as a method of positioning which works at room level accuracy. In nature bats use ultrasound signals to navigate and hunt, this has inspired the design of RTLS that work in a similar manner. Ultrasound signals have a frequency above the human hearing limit (approx. 20 KHz) which enables them to be used without people noticing. Commonly these systems consist of receivers and tags. The receivers are placed at known locations and are simply microphone sensors. The tags are inexpensive and emit an ultrasound pulse which is heard by the receivers. As these signals travel at the speed of sound (343 m/s), Time of Arrival (TOA) to the various sensors may be used and position calculated from this.

The Active Bat positioning system, using ultrasound was designed by AT&T's Cambridge Research centre in the late 90's and was highly successful. While accuracy levels in the order of cm were achieved for 95% of estimates, to achieve this 720 receivers fixed to the ceiling were required to cover a building of 1000 m^2. Thus, the system had low scalability and would not be suitable of widespread use. A more recent ultrasound RTLS is that provided by Sonitor Technologies [11]. A feature of this is that the ultrasound signals are confined to one room and do not pass through walls. These systems need to be combined with RF technology to synchronise and coordinate the receivers with one another. While ultrasound is much less expensive than using IR technology, its accuracy is of the order of centimetre as compared to millimetre for IR. A building wide implementation of an ultrasound positioning system would be extremely time consuming and complex and would therefore have very high overhead associated with it. It would however show how centimetre level accuracy is available if cost is not an issue. Ultrasound systems do suffer from interference for simple noise sources such as crisp packets or jingling keys.

---

[11] http://www.sonitor.com/

8.3 Wireless Sensor Networks – Zigbee

Sensors are commonly used as a means of detecting an environmental or physical condition such as sound, light, pressure or temperature. When a large number of sensors are connected together for means of communication via RF a Wireless Sensor Network (WSN) is formed. Zigbee is a technology standard based on 802.15.4 which allows for control and communication in WSNs. Each sensor in the network is called a node. The nodes form a mesh network and due to the self forming and self healing architecture of WSNs it may be used as an infrastructure for positioning. A Zigbee tag may be localised upon entering the network by taking advantage of the way the network operates. Zigbee routers and tags, periodically or on demand, take TOA values from the signals received from one another. This information may be used by a Zigbee positioning engine to calculate the position of the mobile Zigbee device, given that the positions of the other Zigbee nodes (including routers) have already been calculated relative to one another.

Many of the advantages of using Zigbee are similar to those of UWB systems as the underlying technology is the same. However, their best feature, which is unique to WSNs, is their high fault tolerance. If a network node fails, the network reconfigures and works without it in much the same way the internet does. Despite this significant strength however WSNs suffer from the same weaknesses that all RF technologies do like interference and security issues. The range is short and many nodes are required to give decent accuracy levels. Similarly to many of these proprietary technologies, the installation of a new RF network is unlikely to be popular with I.T. staff and leads to large roll out costs. Also, these new networks need to be troubleshot and maintained by a member of staff. These are costs that may not be apparent when first deploying an RTLS.

8.4 Infra-RED (IR) Tracking Solutions

Infra-Red (IR) based systems usually operate in one room or an open area. This is because the short range of the IR signals does not travel through walls or doors. These systems usually need a direct Line of Sight (LOS) to the target device. Infrared (IR) systems operate by either the user taking some action to highlight their presence to a sensor or the light pulses worn by a user are detected by sensors. With IR technology, due to the fact that light cannot pass through walls, a sensor is required in every room, behind every blocking wall and in every corridor of a facility using this technology.

These IR positioning systems work in a similar manner to RFID systems. Each user wears a tag that periodically emits a beacon containing some unique information about that tag and hence the person carrying the tag. IR sensors on the walls or ceilings detect the tags and give the location. This is usually a network-based positioning system.

Olivetti research developed one of the first indoor positioning systems (ActiveBadge) in the early 90's using this technology. There are very few modern systems using infra-red tracking currently. It has been superseded by the other technologies mentioned earlier.

8.5 Camera Based

By far the best localisation method used by humans is using vision. By simply directing our eyes at a target (within LOS) we can instantly work out where it is and can make a fairly good estimate of the distance from ourselves to the target. Despite the high quality video technology and the powerful computers available today these operations, in general, cannot be performed with the same ease by an artificial vision system. Nevertheless, vision based positioning systems are very powerful if certain criteria are met. By using one or more cameras trained on an area, it is possible to track the location of a person or thing if (a) the computer knows what to look for, (b) conditions for viewing are suitable and (c) the computing engine has enough processing power to perform the complex analysis required for identification and tracking. A problem scenario would be trying to get a computer to spot a previously known person in a crowd who had changed their appearance somewhat. A beard or dyed hair, substantially changes the look of a person in the computers eyes, whereas for humans, the same person would be easily recognisable.

Despite these difficulties, vision based RTLS are rapidly evolving and improving thanks to artificial intelligence techniques and more powerful computing engines. One of the greatest benefits of vision systems is that no tag is necessary. In this way localisation of a person or object doesn't require their cooperation. Locating engines look for specific patterns and can monitor an object in LOS over a large area and over a long timescale given enough processing power. The United Kingdom is reported as having the highest per capita number of surveillance cameras of any nation in the world and is using vision based positioning systems to do it. Setting up an accurate vision based RTLS can be expensive, at least one camera is required per room (preferably more). High network bandwidth is required to transfer a stream of high resolution video between the network of cameras and the positioning engine, which must be a high end computer. These requirements mean that vision based positioning is not a viable solution where affordable ubiquitous localisation is required. If reductions in cost of vision systems can be made, then their readily understandable output and the fact that they do not require tags, could make vision based RTLS a very attractive option. Some of the more recent camera based systems are included here simply for completeness.

8.5.1 Shopper Tracker

An interesting concept which is attempting to provide access to the same kind of analytics available to website admins is Shopper Tracker. Shopper Tracker is developed by Argentinian developer Agile Route. It is built to integrate with the Microsoft Kinect. It analyzes customer movements to provide traffic flow analysis and heat maps indicating which shelves are attracting shoppers and which products they touch or take. This can be tied to conversion data by product SKU to help merchants optimize where products are placed within their stores.

To attain similar market research, merchants typically have to pay observers or use equipment and surveys that are expensive, inaccurate, and influence the behaviour of the people they are studying. With Shopper Tracker, multiple shoppers can be simultaneously tracked around the clock. As it does not cost much more to conduct longer studies, merchants can get more confident results and do A/B testing. There is a demo video of Shopper Tracker in action here[12] and here[13].

Contact: Agile Route - http://www.shoppertrack.com

---

[12] http://vimeo.com/33179742
[13] http://www.shopperception.com/

8.5.2 ShopperTrak

ShopperTrak offers a product which counts people, analyses data, and reveals total sales opportunities for retailers and centre owners. They claim to have over 40,000 managed devices in service and billions of shopper visits counted annually.

ShopperTrak install and configure the hardware, remotely collect traffic and POS data, and format web-based reports for daily review. It is a 100% managed solution. ShopperTrak claims to yield the usual figures such as revenues, how many people will shop, where people will shop and how the market compares to your operation. Again, it is an anonymous service with regards peoples identity.

Contact: http://www.shoppertrak.com

8.5.3 Prism Skylabs

Prism Skylabs is a cloud-based service that allows business owners to bring video feeds online, capture images from these feeds and share this data with consumers and the public. Most stores and restaurants have surveillance videos running 24-hours a day. Unless there is a theft or another crime that takes place in the establishment, this massive amount of surveillance video is unused. That's where Prism Skylabs comes in. A business can download free software that detects cameras or video on a network and showcases a number of images of the space to the business. Similar to the way you can pull images from videos using a video editing software, Prism Skylabs pulls relevant images of your establishment and builds insightful visualizations from these photos, while protecting customer privacy.

For example, Prism Skylabs can extract a visualization that will show the path that people are taking in a store (which can help owners gain insight into the performance of design or display), a heat map of bodies, a photo without any people in the store and more. Owners can gain insight about the performance of the store and the flow at certain times. Prism Skylabs allows users to share and syndicate these photos directly from their platform to a business' Facebook page, Twitter stream, website and Yelp profile. Users can compare web analytics, social media traffic and more with traffic documented by the video images. For example, the startup says that a store could see how much foot and web traffic a Groupon or other daily deal could bring in and at what times the traffic is the greatest. Basically, Prism will be able to mashup data like Google analytics with the intelligence from the actual traffic in the store. The software integrates with point of sale systems as well. Another use case is the ability to post realtime pictures of how crowded a restaurant or bar is in realtime. So a restaurant could post pictures of how empty or full the space is to Twitter or Facebook so potential customers can see how busy an establishment is.

Contact: http://www.prismskylabs.com/

9. Visitor Footfall Capture

Visitor Counting Technology used in conjunction with suitable sensors are used to capture data and count people (or monitor footfall) in retail outlets. By knowing how many people have entered or left each establishment then the effectiveness of sales and marketing campaigns can be monitored. People counting techniques are also useful for monitoring the sales hit rate, i.e. how many sales are made per person entering the store. We cover a number here but others include prodcotech[14]. Most of these include very little details on what they actually offer and how much they charge for their services. Some of the leading solution providers are:

9.1 Experian FootFall

FootFall is a leading provider of information and solutions related to the numbers of people visiting retail outlets, shopping centres and transport locations. Experian FootFall does not manufacture its own proprietary counting equipment. They source from other providers. The counting technologies that Experian FootFall installs are Thermal Cameras, Stereoscopic Video Cameras, Infrared Beams and Road Induction Loops.

Experian globally count more than 6 billion people every year. The aim is to help clients understand their market opportunity through the measurement of customer numbers, and the provision of related metrics, such as conversion rates

Contact: Experian FootFall, Yorke House, Arleston Way, Solihull, B90 4LH
T   +44 (0) 121 711 4652   F   +44 (0) 121 711 8318   W
http://www.footfall.com

---

[14] http://www.prodcotech.com

9.2 IN4MA

IN4MA provide a number of products but they also offer an RFID based system to monitor both customer and staff footfall. Their solution allows footfall data to be remotely viewed centrally, so you could look at many stores across regions to compare effectiveness geographically.

The IN4MA utilises the GSM mobile phone network providing a global solution for remote monitoring. The IN4MA has the ability to integrate into third party devices and equipment - communicating directly with sensors, devices and other forms of equipment. The IN4MA claims to report on changes in state and threshold movements, reporting immediately alarms, which can be sent directly to a monitoring station, a PC, a server, laptop and to a mobile phone via a SMS text message. You can view data and handle alarms using a web browser over the internet. The use of a GSM telemetry device allows real time reporting across the globe using wireless techniques viewing data on their mobile phone, their computer or any Internet capable device.

Contact: sales@in4ma.co.uk
T 01827 310666   http://www.in4ma.co.uk

9.3 CheckCount

Checkpoint Systems offer a web-hosted visitor counting and reporting service. It measures visitor counting data. The 'web' interface' provides trend analysis as well as 'like-for-like' comparisons based on previous periods' performance. Visitor counting data can be sent via email in a format that allows the integration into a sales reporting system. It also allows custom reports to be generated. They claim it is an easy process to integrate their CheckCount hardware system. Collecting data can be done through a LAN via a Global System for Mobile Communications (GSM) network.

Contact: Checkpoint Systems Ltd, Leat House, Overbridge Square, Newbury, Berkshire, RG14 5UX
P: +44 (0) 1635 567070   E: ukinfo@eur.checkpt.com
http://www.checkpointsystems.com

9.4 Axiomatic Technology Ltd

Axiomatic supply compact traffic counters mounted at entry points which emit and detect an infrared beam, counting each interruption caused when a person walks through the beam (see Figure 31). Their accuracy however is reduced by larger pedestrian flows or other factors that can affect the beam, such as wider entrances, inward opening doors or direct sunlight. The units need to be mounted opposite the supplied reflector to count. They offer two beam counter solutions.

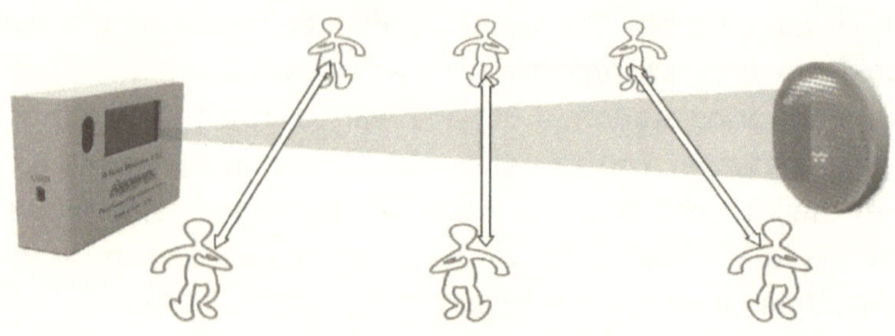

Figure 32: Q-Scan beam counters

9.4.1 Q-Scan Unicomm V2.0 - single beam counter

These are relatively simple to install and come supplied ready to use, complete with mains adaptor, reflector and remote control. You can set them to count either legs or bodies, and automatically divide by two where the entry point is also used as an exit. The unit needs fitting to a solid surface on one side of an entrance, with a reflector mounted directly opposite. It is suitable for single entrances no greater than 6 metres in width.

9.4.2 Q-Scan Twincomm V2.0 Twin Beam Counter

They also supply a bi-directional counter with increased count memory and reporting. Twin beam counters allow counting people moving in a specific direction giving both 'In' and 'Out' counts. Again, this is also installed on a solid surface at an entrance, and is supplied ready for use, complete with mains adaptor, reflector and an infrared remote control. It is suitable for single entrances up to 6 metres in width. It is also available as an Ethernet POE unit that can be networked to our standard Q-Scan manual reporting, automated standard web reporting or their full recap reporting system.

Contact: Axiomatic Technology, Graphic House, Noel Street, Kimberley, Nottingham, NG162NE   **T** 0115 8757505 for more details.

9.5 Euclid

Euclid Elements provide a product that lets shop owners know exactly how many people walk into or even pass by their shops on the sidewalk or in the parking lot. Euclid achieve this by outfitting each shop with a sensor. This sensor looks like a tiny router that Apple would make. It picks up the signal of any smart phone within 60 yards. So long as your phone is Wi-Fi enabled the sensor knows a customer or passer-by is present.

The device then takes the signal, scrambles it and shoots it to Euclid's servers, where the information is sliced and arranged so it is useful. Each shop owner pays about £130 a month for the service and they can then see the data trends on a simple dashboard accessed on a website (see Figure 32). What the storeowner cannot see is any information about a specific customer.

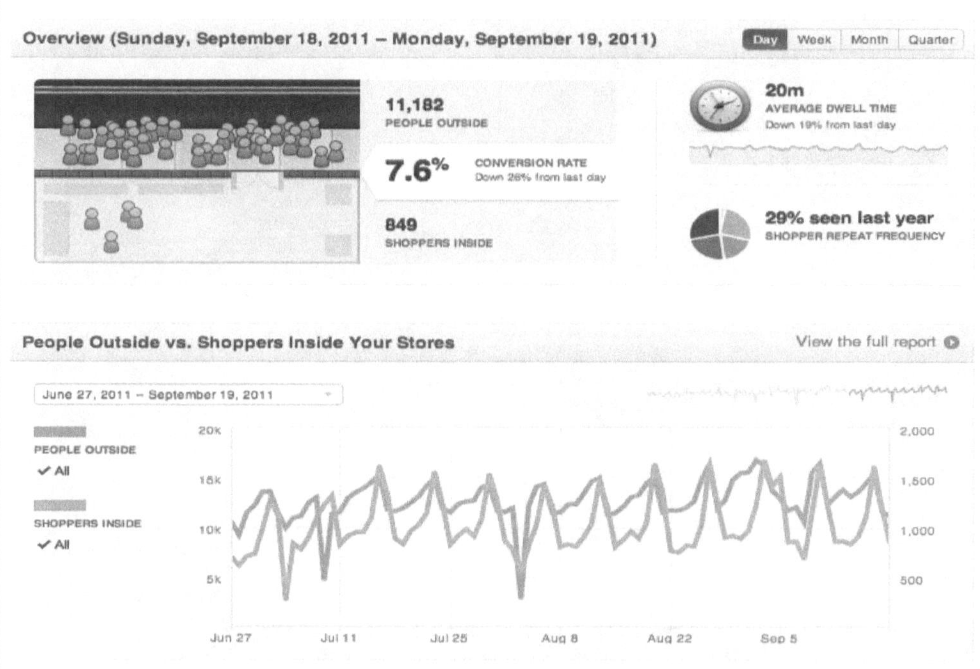

Figure 33: A sample of Euclid's analytics

Euclid have installed sensors in quite a number of stores. One owner found out that customers on average spend 43 minutes in his store in Berkeley, yet only 15 minutes at the one in the SOMA neighbourhood of San Francisco. This allowed him to consider changes such as pulling a sofa from the San Francisco shop to make more room for customers in line. In Berkeley, he could add more food since people are lingering.

He can even figure out the menu items since he knows what time of day people hang out the longest. It is easy to see how shops could do the equivalent of A/B testing. A clothing store might, for instance, switch its window display after discovering that only a tiny fraction of the people who walk by come in - just as e-commerce stores experiment with colours on the home page and track which ones do better.

Contact: info@euclidelements.com   http://www.euclidelements.com

10. Mobile Loyalty

A related area to ascertaining demographics of visitors in the retail sector is mobile advertising and mobile loyalty. This set to rapidly grow in the time ahead (e.g. Google just spent $750 million buying a mobile ad platform – adMob which is especially popular on the iPhone). Loyalty marketing programs are designed to increase customer satisfaction and retention through communications-based loyalty and rewards programs.

Path Intelligence seem to be about to release a mobile coupon service called Flockr (http://www.goflockr.com) but little seems to be known about it apart from what they claim on the site which seems to be a way to offer opted-in customers a method of getting discounts when certain shops are quite. Starbucks is running a loyalty program based on 2D bar-code coupons deployed via SMS. The coffee giant tapped digital technology provider Codilink, which specializes in mobile services helping companies to offer loyalty programs to their clients. The campaign is currently running in Guadalajara and San Luis Potosi, Mexico. IKEA has launched a mobile loyalty program to build a database of consumers interested in receiving discounts from the home furnishings retailer. It is a "text-to-enter" type loyalty program and it is meant for IKEA to start a dialogue with interested consumers. BCode is powering the texted discounts. Another example of Mobile Loyalty is from Sprint Nextel and the Luxor Hotel in Las Vegas for a soon-too-launch campaign. When a guest makes a reservation at the Luxor, they were asked if they wanted to opt-in to receive automatic check-in when they arrived in Las Vegas. Then, using location services, the consumer received a text message when they arrived in the Las Vegas city limits prompting them for automatic check-in via their mobile device. For those who have not travelled to Las Vegas, the queues for check-in are often enormous. This new service enables the consumer to bypass the check-in lines when they arrive at the hotel. This is an example of offering an incentive big enough to get the customer to buy in and allow tracking.

DDR Corp., which owns 500 outdoor shopping centres across United States, Puerto Rico and Brazil, is texting customer's coupons as they approach the parking lot. Many retailers are employing the technology in an effort to discourage shoppers from using their smart phones to compare the price of goods they see in the aisles of one store with better deals that may be available elsewhere. The technology is being used by other kinds of related businesses—including grocery chains and retail landlords such as DDR. DDR is conducting a six-month test of ValuText, a marketing tool that sends offers to customer cell phones when they are close to stores. The 27-shopping center test, which started in December, is an effort to fight back against online shopping which has taken business away from brick-and-mortar retailers across the country. DDR's malls are usually anchored by retailers like Wal-Mart, Bed Bath & Beyond and Kohl's. Such big box retailers have been hard hit in recent years by the recession and the rise of online shopping.

DDR remains unprofitable after the recession and the rise of online shopping pushed major tenants like Linens 'N' Things, Circuit City and Goody's into bankruptcy. The technology, known as geo-fencing is an attempt to draw customers back into the physical stores DDR depends on for its business. The strategy is also a response to customers who already come into stores but treat them as showrooms, ordering online after they are done browsing, or using mobile phones to search for cheaper prices in the stores of competitors. Placecast, the company which created the geo-fencing technology used by DDR, has deals with major cell carriers to periodically check if customers are within the designated area. When customers pass through this invisible perimeter they receive deals–20% off, two-for-ones–the same kind of offers found in traditional coupons. About 10% of the 7,000 consumers who signed up to receive the offers have used the coupons. So far the program is not very sophisticated in its targeting of customers. DDR does not have access to sales information for the stores, so they cannot target customers based on past purchases.

10.1 Shopkick

Shopkick is a shopping rewards program. When customers with their app walk into convenience stores, they automatically rack up instant über-versatile points called "kicks." It is a free app on iPhone and Android. If customers link their credit cards, then they collect more kicks for purchases. These kicks can be used for treats such as a free Starbucks latte, Facebook credits, iTunes downloads or actual products and discounts.

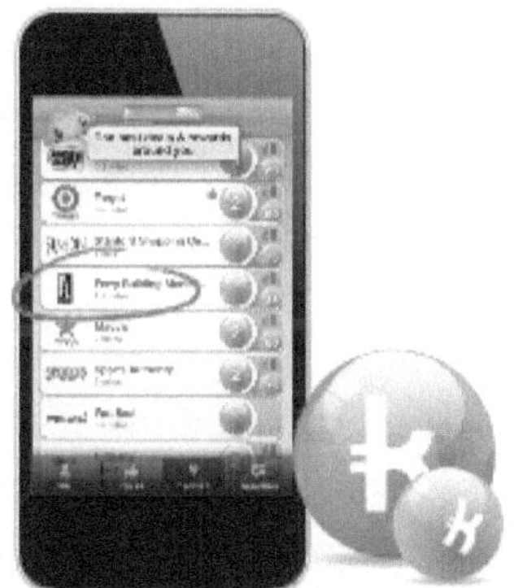

1. Consumers download the free app and look for interesting stores and offers.

2. Consumers find nearby stores and restaurants with great products and kicks rewards.

3. Consumers walk into your business, get rewarded just for visiting and make a purchase.

Figure 34: How Shopkick works

Apparently, the shopkick app is already one of the 5 most widely used shopping apps in the country, according to Nielsen, along with giants like eBay and Amazon. It is even more used than any physical retailer's own app. It is used in over 7,000 stores across America. Over 3 million people have used the app. The company remains private so financials are unknown. This concept does show how mobile traffic may be monetized. In this case, customers seem attracted to locations where they can use coupons and collect kicks for gift cards. The future potential still exists for locations to alert or contact nearby customers, but that theory still seems questionable

Contact: partners@shopkick.com or http://www.shopkick.com

10.2 Near Field Communications (NFC)

Near Field Communication (NFC) is a technology that enables a device to communicate with another at a maximum distance of around 20cm or less. Currently, mobile phone manufacturers, banking institutions and mobile network providers are attempting to apply this technology to Smartphones and other handheld devices because of the opportunity to enable the consumer to use commercial services more easily.

As more phone manufacturers start to include NFC chips in their mobiles, the need for applications will increase. Already marketers are looking at the possibilities of using the NFC interface alongside their traditional marketing methods such as posters. Information could also be passed to the NFC device, allowing the user to gain more information about a product or service, so this would be an efficient means of advertising. For example, it would be possible to transmit a URL to the target device so that the user would then be able to navigate to a website to get further information about a product or service in which they are interested. This is where having NFC enabled on a smartphone could prove to be very useful for consumers, enabling them to find out the best price for a product before committing to the purchase. There are many uses for NFC. They can also be used to transfer tokens at airports, which would eliminate the need for boarding cards. The passenger would check-in using their mobile and then re-confirm by swiping their phone again at the departure gate. There is also the possibility of them being able to store biometric information, which is becoming more widely developed for security.

NFC devices can be used in conjunction with image display devices like digital photo frames for displaying images very quickly. All the user needs to do is touch the photo frame with the image ready to be sent, then the connection is established and the image is sent over Bluetooth. NFC is backward compatible with RFID therefore it is perfectly feasible to use an NFC enabled device as an RFID key. This can be used with traditional RFID access control systems as a replacement for the key fobs and cards currently used. Wireless car keys using NFC are being developed by BMW with personalised settings stored into each key. They have developed an NFC car key system which will link into the cars current navigation system which already allows for hotel reservation, and train ticket booking. Using NFC the tickets and reservations can now be stored on the NFC card which can then in turn be used to gain access to the hotel room or validate the ticket with the conductor. Applications for smartphones are starting to appear that allow the user to create their own NFC tags, an application that was developed and is being distributed for free is NXP TagWriter for the Android smartphone. The application uses the NFC enabled phone to send a signal to write contact details, URLs and SMS messages onto an NFC enabled tag which can be on items like business cards up to posters.

10.2.1   NFC and Windows 8

Microsoft's upcoming Windows 8 operating system (OS) will include built-in NFC functionality. Although Microsoft has yet to set a date for the product's release, the company reports that the new system will include an NFC function known as *"tap to share,"* enabling Windows 8 PCs, laptops or tablets to support NFC RFID readers. In that way, the firm indicates the computing world will join a limited number of mobile phones that are NFC-compatible, acting as 13.56 MHz passive NFC readers and writers that can interrogate tags and capture as well as send data wirelessly when within range of those tags. Microsoft have recently released a Developer Preview build of Windows 8, known as Build 8102, for software and hardware developers to download and begin working with.

The tap-to-share application includes software and driver files to enable the use of a plugged- or built-in NFC reader in order to receive or transmit information to or from another device, such as an NFC tag, an NFC-enabled phone or another NFC-enabled computer running Windows 8. With Windows 7, on the other hand, a user can connect an NFC reader to a computer or laptop, but additional software and a driver, supplied by the reader manufacturer or a third party, are necessary to capture and interpret data transmitted to that reader. During its Build conference, Microsoft demonstrated the tap-to-share application by means of an NFC-enabled tablet computer loaded with an early version of Windows 8. In response to Microsoft's new OS plans, chip manufacturer NXP Semiconductors has announced that its PN544 NFC radio controller is compatible with the new operating system. In fact, NXP provided the NFC technology used on Windows 8-based tablets distributed at the conference, enabling the computers to not only read and encode NFC RFID tags, but also support peer-to-peer and card-emulation functions specified by NFC standards developed by the NFC Forum.

In October 2011, RFID tag and inlay manufacturer UPM RFID announced that it has teamed with NFC solutions provider Wireless Sensor Technologies (WST) to provide NFC tags. With this partnership, UPM RFID is providing WST with its NFC tags and inlays that will now be added to WST's NFC readers, software development kits and customization services, including printing and encoding. Although the partnership is not a direct response to the Windows 8 NFC plans, WST notes, it will make it easier for the firm to respond quickly to customers seeking technology solutions built for Windows 8 devices. Wireless Sensor Technologies already provides an NFC app known as GoToTags for computers running Windows 7, enabling users to communicate with an NFC reader plugged into their computer and to use it to read and encode NFC tags.

Those in the NFC industry claim that the tap-to-share application signifies that the use of Near Field Communication will grow as NFC support in Windows 8 should spur the new community of developers and end-product manufacturers to create new applications. NFC's use in personal devices which may now include tablets and laptops, as well as mobile phones should enable brick-and-mortar stores to link their products with the Internet.

10.2.2   Google Wallet

Google Wallet[15] is an Android app that makes your phone your wallet. It is primarily aimed at the payments market as it stores virtual versions of your existing plastic cards on your phone. It works however by people tapping their phone to pay and redeem offers using near field communication (NFC). It is just being rolled out around the world. Google Wallet has been designed for an open commerce ecosystem. It aims to eventually hold many cards people keep in their leather wallet today. Because Google Wallet is a mobile app, it will be able to do more than a regular wallet ever could, like storing thousands of payment cards and Google Offers but without the bulk. Google hope that eventually our loyalty cards, gift cards, receipts, boarding passes, tickets, even our keys will be seamlessly synced to our Google Wallet. Every offer and loyalty point will be redeemed automatically with a single tap via NFC. The vast majority of phones however do not support NFC but Google believe that NFC will be surging in popularity over the next couple of years, and for the time being this is really a first step. Google also has a plan to enable older devices to use a more limited version of the app - stickers that you can put on the back of your phone.

Google are a little vague to date on this but it seems the plan is that users will be able to obtain special NFC stickers with a single credit card associated with them (such stickers already exist, but these stickers will apparently be able to communicate with the Google Wallet app). Transactions made using the sticker will be relayed to the Wallet application on an Android device via the cloud. It is possible this functionality will be extended to other platforms as well, as Google says it is willing to partner with everyone to help broaden support for Google Wallet. Google Wallet is now released on the Nexus S 4G by Google. It is possible that there may be real opportunities in the mobile tracking of visitors arena to use the NFC on the Google phones to provide added value and services. The obvious feature is the knowledge of the location the user is at and the fact that NFC should be deployed on most smartphones in the near future. It does however have a very short range.

---

[15] http://www.google.com/wallet/

10.2.3   NFC Security Issues

It is estimated that the market in NFC devices will grow exponentially over the next few years, and with this comes the always-present issue of security. The requirement for the two devices to be close to each other is something that helps with the security of the transaction by limiting eavesdropping. The range of NFC is only a few centimetres. This makes it inherently safer than longer range technologies but there are still security flaws that, if not addressed, can be exploited. According to the ISO standard, NFC is not encrypted. This is to make it backward compatible with RFID technologies. Encryption may be implemented with future NFC applications but only as a best practice, not as a requirement. The wireless signal generated by data transfers can be picked up by antennas, modified, and dispatched. This makes NFC inherently vulnerable to this kind of attack. In active mode where the two devices are communicating, eavesdropping is significantly easier compared to passive mode as the antenna signal gain is lower, therefore the listening range is greatly decreased. It has been found that when in it is easy to successfully eavesdrop from around 30cm. The AES encryption method is developed into a series of NFC security standards to protect against eavesdropping and data manipulation. An NFC skimmer device, similar to the magnetic strip skimmer used in ATM machines could be possible to implement. With a disguised device placed close to the two NFC devices, it would be able to record all NFC activity in a given time and be collected at a later date. NFC is starting to become popular as a form of advertising, where the interested user taps their device onto the advert to view the message, URL or phone number. Fraudsters can take advantage of NFC tags in public places by removing the legitimate tag and replacing it with a tag directing the user to a bogus website of a premium number set up to the fraudsters' account.

Another aspect to protect against is 'Walk off'. Walk offs are when the device user lifts the device and walks away from the transaction while leaving the transaction connection open. Usually, when the connections are idle for an amount of time the connection terminates automatically, but the time window where the connection is still open can be exploited. A wireless key can be used to encrypt the data. Also a new concept is 'Electronic Leash' which terminates the connection once the device senses a set distance has been exceeded. Devices using NFC are expected to operate in environments with varying security, some with a high level of security and others that do not need any security. As the NFC Forum has repeatedly stated the technology is 'inherently secure' because of the small transmission distance. The best solution to these security issues is to have a layered security model with a minimum requirement of authentication before the start of communication. Developers can then add higher levels of security according to their application needs. This is not required by the ISO standard but will be essential for making money from the technology.

11. Conclusion

It seems there is a key distinction between "customers" and "shoppers". This is key to the entire idea of shopper marketing: a customer is someone who has purchased something before (and who may or may not do so again in the future). A shopper, on the other hand, is someone who is actually engaged in the action of shopping, whether by looking through the catalogue, browsing the website, or roaming the aisles of the brick-and-mortar store. Ideally, the right messaging helps convert shoppers into (repeat) customers.

There is a growing need by retail to know who the shopper is, what customer segment they belong to, their transaction history, browsing pattern in shops, which sections they visit and where they spend the most time. Sites such as Amazon do this of course and the holy grail of physical shopping would be to replicate. Enter Presence Marketing. Presence Marketing hinges on two fundamental in-store concepts: Presence Identification and Session Metrics. Presence Identification identifies a customer upon entry to retail establishment and Session Metrics captures the duration of a customer's shopping visit, departments visited, time spent in the various departments. It provides key insights such as lost sales -- when a shopper spends time in a department/store and does not purchase. Using Active RFID with a combination of mobile technology is the appropriate platform to enable Presence Marketing. A loyalty or credit card embedded with Active RFID technology would allow a card carrying customer to be identified the moment they enter a retail store. This would be done with the permission of the customer on an opt-in basis to address any privacy concerns. Active RFID technology would also allow the capture mechanism of the in-store "session metrics" – the browsing pattern of a shopper analogous to the navigation pattern in an on-line e-commerce site. The delivery mechanism for brand communication would be the mobile platform.

Every location determination technology has its advantages and disadvantages in a number of areas namely; if they are designed to operate inside or outside, how they determine their position internally or via a network connection, their cost, their susceptibility to interference and their location determination accuracy. This report has provided an overview into the various RFID, Mobile, Ultra wideband, Map based and Wi-Fi location determination methods and technologies and their uses and sought to provide an extensive and up-to-date market review of available position- sensing technologies ideally suited for tracking customers in retail environments. It could also be stated that the non-802.11 location tracking systems include ones such as ultra-wideband, active RFID, ultrasound and other RF-based systems for the most part require installation of proprietary and single-purpose antennas, and dedicated staff to deploy, manage and maintain. Some of these technologies such as active RFID, are problematic in certain scenarios because they can introduce possible interference with Wi-Fi networks and critical patient care equipment. 125 kHz chokepoints have been banned by many hospitals around the world because they can interfere with clinical equipment in hospitals and ZigBee can adversely impact a customer's existing Wi-Fi network that is used for primary voice and data communications. These aspects are not to be underestimated and may indeed strengthen the hand of the 802.11 location tracking systems in the future.

To successfully deploy a positioning or tracking system based on 802.11 WLAN, some aspects must be considered and planned carefully such as the number of access points. It is important to note that more access points do not enhance coverage. In relation to this, the locations of access points should be strategic. The distance between two adjacent calibrated locations should not be too large ~ 1-2 meters is fine and each location should have enough calibration samples (e.g. 200 to 300 samples). It can be important to give denser calibration locations to the areas which may be confused with other areas and finally to ensure that during the calibration/fingerprinting process that one walks slowly, stopping regularly for up to 30 seconds for increased accuracy. Technologies such as provided by Qubulus and Path Intelligences 'flockr' may have potential but sadly it is not yet available and seems to be going down the route of mobile coupons but Joe reassured me that other layers are being added which will identify customers.

Other players on the edge include CSR[16] which are a UK technology company who specialise in wireless hardware. Chipsets they have designed are incorporated into thousands of consumer electronics products including cars, laptops, games consoles, digital cameras and, of course, mobile phones. Whilst traditionally focusing on such things as Bluetooth and audio streaming, they have recently made advances in the indoor location world, demoing their latest technology at Mobile World Congress in Barcelona. Their approach centres around a new chip which gathers real-time information not only from GPS signals, but also a whole range of other satellite and radio signals along with device sensors to such as accelerometers and gyroscopes. This data is

---

[16] http://www.csr.com/

then combined with external data such as mapping and WiFi hotspot location databases to come up with a form of "dead reckoning", sufficient for the device to calculate its position indoors. Certainly one to watch as if these chips start to make it into the next wave of smartphones, and work as well as CSR say they do, indoor location could quickly go mainstream. However, their systems are not something that can be leveraged right now for location estimation in retail.

Pole Star[17] based in France has developed their indoor positioning system based on existing WiFi infrastructure on-site. Their model appears to be to sell the solution to site-owners, allowing them to map their buildings and then make the information available to apps that implement their system. The offering is a complete service of coming out and taking readings round the building, creating a 3d map and preparing the whole thing to be inserted into an app. There is a link to apply to become a registered distributor, but no downloadable software for developers to experiment with. They claim it will work on iOS, Android and Windows Mobile.

It seems that for now, Active RFID solutions are the most cost-effective way in which to track and identify shoppers in real-time in indoor locations but we expect new technologies as offered by Path Intelligence and Qubulus and Sensewhere such as outlined here to supersede Active RFID in the not too distant future.

---

[17] http://www.polestar.eu/en/

Appendix A: RFID Presence Marketing Typical Deployment

Summary

A retail client wish to track the location of people in public locations - most likely a shopping centre. If they were to deploy a series of Active RFID Readers/antennas - then there would be a limit as to how many can be deployed (due to price). To track three individuals locations such as shown in Figure 34 would result in a portal being placed at the entrance to the centre and then at each door to the shops in order to ascertain visits and dwell time. Here we are presuming that each shop has only one single entrance/exit.

Locations

A Shopping Centre

Physical Scenario

The initial trial could take place in a scenario similar to Figure 34. This is where the people could be tracked. Each shopping centre deploying an active RFID or wireless 802.11 or mobile phone tracking system will have to be on board however as access to the infrastructure is required.

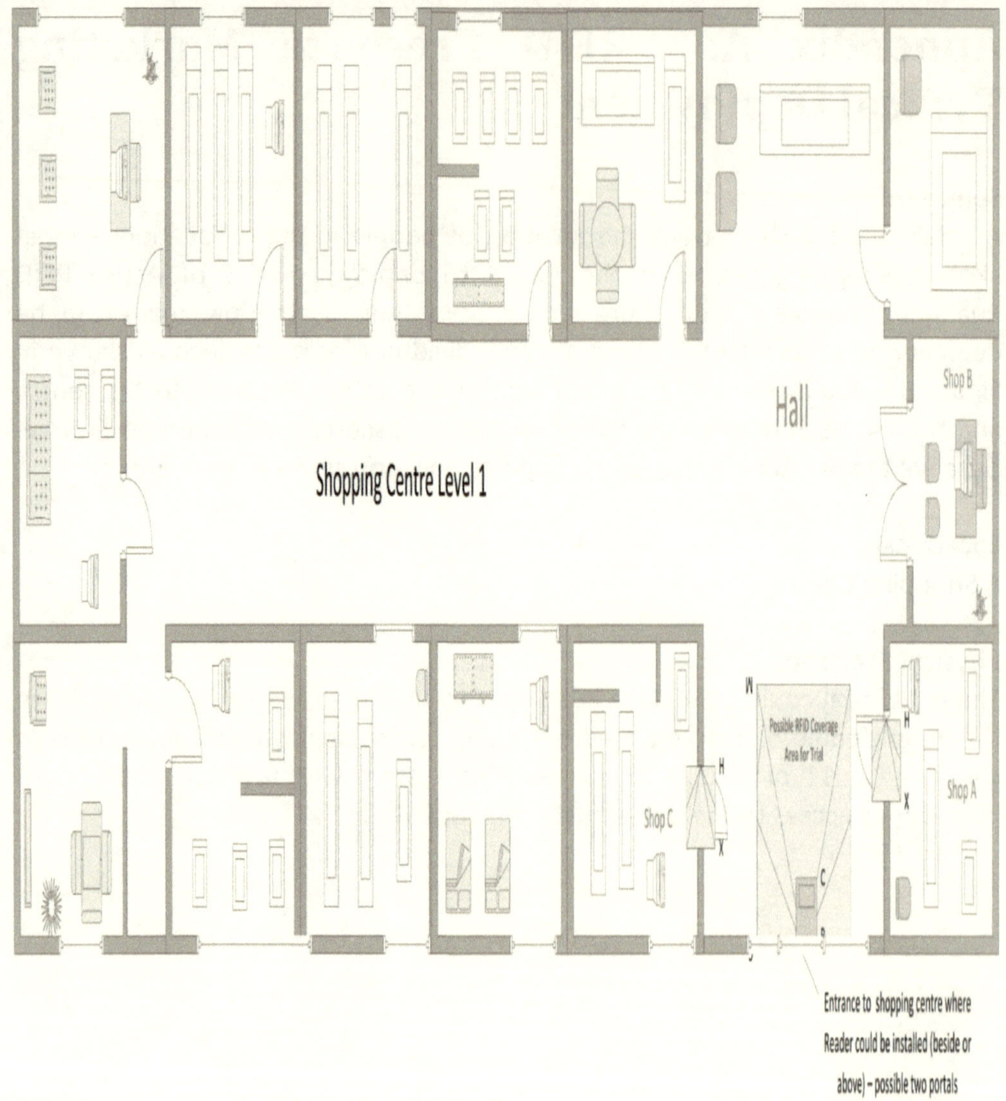

Figure 35: Layout of Shopping Centre Retail Units for Initial Feasibility Study

Appendix B: WiFi Presence Marketing Typical Deployment

With all the typical Wireless 802.11 positioning systems outlined earlier, an initial fingerprinting phase will occur where staff involved in the deployment will walk around the shopping centre with plans of the building loaded on a laptop.

A recording of 'wireless fingerprints' will then be saved. This allows us to locate people in the live scenario. Basically, a database of fingerprinting will be calibrated to determine the location of people when they are within the range of the AP's, so almost any person within the centre can be located. These people will either be tracked using their mobile phone (WiFi turned on) or the specific WiFi tags purchased.

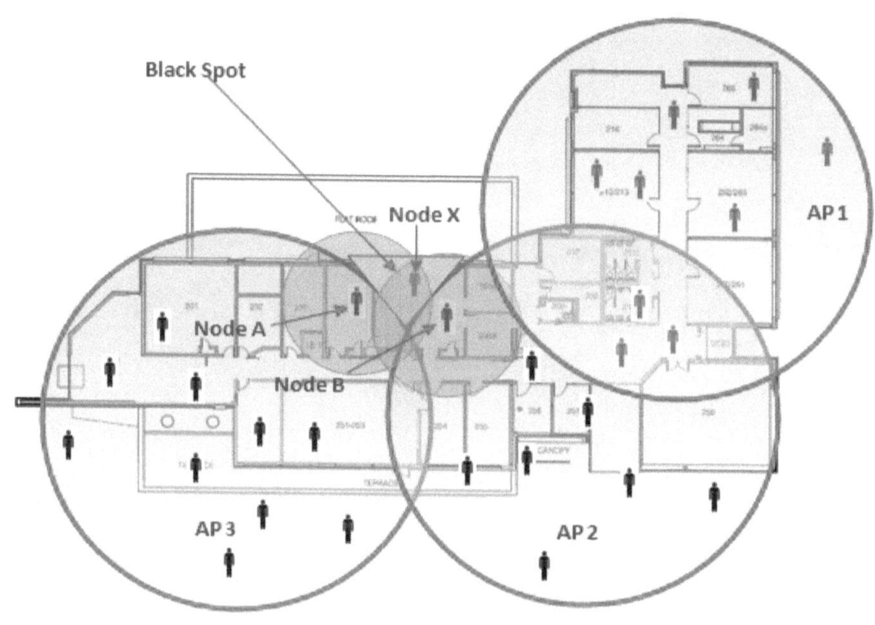

Figure 36: Fictional L shaped shopping centre floor with Access Points placed to cover most of the floor area

For example, Figure 35 illustrates a section of a typical shopping centre with Wireless Access Points strategically placed to cover as much of the 'L' shaped building as possible, given the limitations of the range of the devices. The circular shape outlines the coverage of a number of AP's and is used to illustrate the coverage area and the Black Spot, but should not be confused with a diagram of only 3 AP's. The 'Yellow' triangle shaped area at the front of the building, labelled as 'Black Spot' is the one area of the building that is not covered by an AP, therefore person X's position cannot be located using the AP's. Person X will be referred to as the 'Blind' Person as we cannot ascertain their location at this stage. Person A and Person B are located at the outer reaches of the AP's but their positions are 'known'.

This scenario is mainly to demonstrate that there will be blind spots. It is technically possible to fix all blind spots by deploying more access points but this can be cost prohibitive. Strategically maximising the positions of existing access points can be more cost effective.